THE DRAGON IN THE COCKPIT

Ashgate Studies in Human Factors for Flight Operations

Ashgate Studies in Human Factors for Flight Operations is a series dedicated to publishing high-quality monographs and edited volumes which contribute to the objective of improving the safe and efficient operation of aircraft.

The series will achieve this by disseminating new theoretical and empirical research from specialists in all relevant fields of aviation human factors. Its foundation will be in applied psychology, presenting new developments and applications in such established fields as CRM, SA and decision-making. It will also encompass many other crucial areas such as fatigue and stress, the social environment, SMS, design, technology, communication and training.

Submitted work relevant to the objective of the series will be considered for publication by the board of editors. The series is intended for an international readership and so books with a broad geographical appeal are especially encouraged.

The Dragon in the Cockpit

How Western Aviation Concepts Conflict with Chinese Value Systems

HUNG-SYING JING
National Cheng Kung University, Taiwan, Republic of China

ALLEN BATTEAU
Wayne State University, USA

ASHGATE

Published by
Ashgate Publishing Limited
Wey Court East
Union Road
Farnham
Surrey, GU9 7PT
England

Ashgate Publishing Company
110 Cherry Street
Suite 3-1
Burlington, VT 05401-3818
USA

www.ashgate.com

British Library Cataloguing in Publication Data
A catalogue record for this book is available from the British Library

The Library of Congress has cataloged the printed edition as follows:
Jing, Hongxin, author.
The dragon in the cockpit : how Western aviation concepts conflict with Chinese value systems / by Hung-Sying Jing and Allen Batteau.
pages cm -- (Ashgate studies in human factors for flight operations)
Includes bibliographical references and index.
ISBN 978-1-4724-1030-6 (hardback) – ISBN 978-1-4724-1031-3 (ebook) – ISBN 978-1-4724-1032-0 (epub) 1. Airplanes—Piloting—Human factors—China. 2. Intercultural communication—China. 3. Aircraft accidents—Human factors—China. 4. Aeronautics—Human factors. 5. Aeronautics—Safety measures. I. Batteau, Allen, 1946– author. II. Title. III. Series: Ashgate studies in human factors for flight operations.
TL553.6.J56 2014
629.132'520951--dc23

2014023412

ISBN 9781472410306 (hbk)
ISBN 9781472410313 (ebk – PDF)
ISBN 9781472410320 (ebk – ePUB)

MIX
Paper from responsible sources
FSC www.fsc.org FSC® C013985

Printed in the United Kingdom by Henry Ling Limited, at the Dorset Press, Dorchester, DT1 1HD

Contents

List of Figures

List of Tables

Preface

In the past 30 years, China has slipped the surly bonds of its imperial past and joined the modern world economy, building sophisticated industries in electronics, pharmaceuticals, information technology, and most recently commercial aircraft. The COMAC C119, China's bid to join the United States and the European Union as exporters of wide-body commercial aircraft, is but the most recent of three decades of bold moves where China is re-asserting itself as *Chung Kuo*, the Middle Kingdom. However, in today's world of large technological systems such as air transport, global supply chains, and the internet, this creates challenges that would have been unfamiliar to travelers in the past, whether Marco Polo in 1271 or Lord Macartney in 1793.

The Dragon in the Cockpit examines one of these challenges, the inevitable and sometimes disastrous collisions of technology, language, and cosmology created when an advanced technology created in the West is adopted in the East. The consequences of these collisions are well known: In North America and Europe, there are fewer than one fatal accident per million flight departures, whereas in China there are 3.8 fatal accidents per million departures. Despite the fact that Chinese history is marked by many of the technological advances that shook the world, including gunpowder and the compass, this mismatch between East and West takes on a new aspect when embodied in large-scale, high-performance technologies. Nor is this simply a matter of China "catching up" with the West: the relentless, geometric acceleration of technological advance creates a Malthusian race between technological invention and technological adoption around the world, which will assure that there are always global mismatches between technology and culture, often with tragic consequences (Batteau 2010).

In this book we provide a new perspective on the differential successes of Chinese and Western civil aviation, in the contrasting cosmologies and value systems of China and the West. The Western cultures that invented the airplane and created global air transport networks are individualistic, egalitarian, and linear in their thinking. By contrast, Chinese culture is relationship-oriented, hierarchic, and figurative. These differences at times have tragic consequences. As currently designed, a modern airliner is a crew-served device, and flying it safely depends on a high degree of coordination among the flight crew. This is so important that an entire training discipline, "Crew Resource Management" teaches assertiveness, task-sharing and other communication skills to flight crews (Chapter 6). Assertiveness, however, is a foreign concept in a hierarchic culture, and several accidents have resulted from junior officers failing to correct obvious errors by their seniors (Chapters 1 and 2).

The situation becomes even more complicated with the addition of flight management systems to virtually all modern airliners. These systems are essentially computers that fly the airplane, and their programming embodies the logic of the Western engineers who designed them (Chapter 3). The complexities created by these "third crew members" have been implicated in several accidents among European and North American airlines; when one adds the cultural and linguistic differences between East and West, the sources of misunderstanding skyrocket.

This book had its origin in the decision by Jing Hung-Sying in the 1990s to dedicate his career to interpreting Western technology in the Chinese context, and to a very specific event in September, 2001. Jing, his wife, and Batteau were having dinner in a restaurant in Tainan when they received word of an event halfway around the world. It was evening in Tainan, but it was morning of the same day in New York City, where an airliner had just flown into the World Trade Center. Jing and Batteau had been discussing flight safety, and it slowly dawned that the unfolding events of September 11, 2001 were not the sort of aviation accidents that they had been discussing. Over the next few days, while waiting for flights into North American airspace to resume, Jing and Batteau had several opportunities to reflect on and discuss the meaning of these events for the world and for aviation. When Jing asked Batteau to assist with the English translation of his original 龍在座艙 published (in Chinese) in 2009, Batteau readily agreed.

Hung-Sying Jing would like to express his sincere appreciations to Dr. Kay Yong, former director of the Aviation Safety Council for his support of the research project which the present book is based, and to his colleague, Professor P.-J. Lu as the principal investigator of this research, and Thomas Wang for his full involvement in the survey work, and to his former student First Officer Casper Wang in China Airline providing valuable insight about CRM, and Shirley Tzeng for her assistance. Special thanks also go to his daughter Catherine Jing, a Columbia University graduate for helping translating the book. Also, without the help from many respected figures in the Taiwanese aviation community, this book would not be possible. They deserve deep appreciations from the authors.

Allen Batteau would like to extend his thanks to Dr. Kay Yong, formerly of the Aviation Safety Council, introducing him to the world of Chinese civil aviation, and to Capt. Rafael Alejandro Pérez Chávez and Dr. Carolyn Psenka, for their involvement in the survey of CRM attitudes among Chinese pilots, and to Suzanne Walsh, for her valuable comments on the final manuscript, and to Dr. Maria Roti for assistance with the bibliography.

The stubborn facts of higher accident rates among Chinese carriers, both in the Peoples Republic of China and the Republic of China (Taiwan), indicate that transplanting this modern technology into the Chinese context faces some challenges. At the same time, the ascent of Chinese civil aviation into a global sky is an expectable step in the technological evolution of the human race, with potentially mutually beneficial results for East and West. Although the equipment is the same, the cultures are different. Examining this cultural difference from the perspective of 6,000 years of history is the objective of this book.

Chapter 1
Tragedy in Nagoya

"Be careful! Guard and cover the memorial tablet! The souls cannot be touched by the sunlight!"

A high and clear voice appeared in Taipei Chiang-Kai-Shek (CKS) International Airport (CKS), as airport officials brought along a group of big black umbrellas (*United Daily News*, April 29, 1994). It was 3 o'clock in the afternoon, the sun dazzled in the sky. There was one victim each covered under the Each victim was covered by a big black umbrella, and with a heartbroken family behind. Just a few days earlier, a plane crashed in Nagoya Airport, Japan. The cremated remains of 67 victims were carried back to Taipei together with 30 coffins, and the reception site for the dead was set up right in the airport. Airline company staff held high a white long cloth, waving to and fro against the wind, written on it with large black characters: "China Airlines staff begged for forgiveness from victims' families," bowed and apologized with deep sorrow and regret again and again …

On April 26, 1994, two Taiwanese pilots flew an Airbus A300-600R automatic airplane from CKS International Airport to Nagoya, Japan. There were 256 passengers, 13 cabin members and two pilots, 271 people in total on the plane. The plane took off smoothly from CKS Airport on schedule; two pilots followed the designated route according to the plan, heading to Nagoya Airport, Japan.

At 8:07pm (20:07), the airplane prepared to land on runway 34 under guidance from the approach tower and the instrument landing system (ILS). At 8:12pm, the plane passed the outer beacon with altitude roughly above 2,000 feet. The plane followed the three-degree glide path marked by the ILS to execute the standard landing procedure. Everything was normal. Two minutes later, when the plane was flying just below 1,000 feet, it started to level off, as if it didn't want to land. Yet, no additional sign showed that the plane was starting to climb to go around, which is a standard maneuver in a missed landing. After 15 seconds of level flight, the plane started to descend. At this moment, the plane was only 800 feet above the ground, and the speed was only 146 knots, yet well above the stall speed of 110 knots.

The plane descended steadily again, and everything seemed back to normal. At 8:15pm, the plane continued descending for about 30 seconds, reaching 350 feet above the ground. It looked like the giant was about to land. Yet suddenly, a deep and loud roar came out from the engine, the thrust was suddenly elevated to the maximum. The nose started to level off, and the plane surprisingly started to climb with increasing speed. It happens once in a while that a plane cannot land smoothly in the very final stage of landing, and has to execute a "go-around," a standard maneuver. However, it did not look like this plane was going around, because the pitch of the plane was still increasing, eventually to 50 degrees!

The wing stalled, the lift dropped drastically, and the drag increased significantly. Unfortunately, the thrust from the engines was not enough to raise the plane by itself. A chain of abnormal events combined and pushed the altitude back to 1,500 feet, with the speed suddenly declining to only 87 knots. The speed was already significantly lower than the stall speed. The plane started to fall out of the sky.

In less than one and half minutes, the situation facing the plane evolved from a three-degree steady descent on landing procedure to an irreversible situation beyond human control. The aircraft lost lift, its drag increased, its thrust was less than normal, its altitude increased, and its speed dropped. The entire situation made the plane totally out of control.

Two years later, the Japanese authority released the accident report (Aircraft Accident Investigation Commission, Ministry of Transport, Japan, 1996). The report pointed out that there was no clear sign of mechanical malfunction or special meteorological factors. In other words, the accident could only be attributed to human factors of the flight crew.

According to the accident chain theory from Boeing (Boeing Commercial Airplane Group, 1994), the Ministry of Transport of Japan listed every single link showing the evolving sequence in the accident:

1. The first officer (F/O) inadvertently triggered the go lever activating the go-around procedures.
2. The F/O continued pushing the control wheel in accordance with the instructions of the captain (CAP) in order to continue the approach.
3. The movement of the trimmable horizontal stabilizer (controlled by the computer) conflicted with that of the elevator (controlled by the crew), causing an abnormal out-of-trim situation.
4. There was no warning and recognition function to alert the crew directly and actively to the onset of the abnormal out-of-trim condition.
5. The CAP and F/O did not sufficiently understand the flight director mode change and the autopilot override function.
6. The CAP's judgment of the flight situation while continuing approach was inadequate, control takeover was delayed and appropriate actions were not taken.
7. The alpha-floor function was activated. This was incompatible with the abnormal out-of-trim situation, and generated a large pitch-up moment. This narrowed the range of selection for recovery operations and reduced the time allowance for such operations.
8. The CAP's and F/O's awareness of the flight conditions, after the CAP took over the controls and during their recovery operation, was inadequate respectively.
9. Crew coordination between the CAP and F/O was inadequate.

The alpha-floor function is an automatic protection to prevent the plane from stalling. When the angle of attack of the plane reaches 11.5 degrees, close to stall, the flight control system will automatically open the throttle to the limit and lock, pushing the thrust to the maximum and speeding up the plane, so as to escape from the danger of stall.

The accident report made some recommendations to Airbus and the airline:

To Airbus:

1. Improvement of the override function, out-of-trim prevention and warning and recognition functions for the trimmable horizontal stabilizer of the automatic flight system of the A300-600R is required.
2. The descriptions of override, disengagement of go-around mode and recovery procedures from out-of-trim in the Flight Crew Operation Manual of the A300-600R should be improved from the operational viewpoint.
3. In the event of an accident or serious incident, Airbus Industrie should promptly disseminate the systematic explanation of its technical background to each operator, and furthermore should positively and promptly develop modifications.

To the airline:

1. Reinforcement of education and training programs for flight crews;
2. Establishment of appropriate task sharing;
3. Improvement of crew coordination;
4. Establishment of standardization of flight.

Stated simply, this accident was caused by the conflict between the pilots and the automatic flight control system. The flight crew were fighting for control over the plane! In other words, the pilot pushed the plane to land, the computer managed to go around, and both wouldn't let go.

In fact, similar incidents had taken place before, incidents that weren't severe enough to cause planes to crash. Airbus has also released a notice, warning the pilots not to try to override the automatic flight control system. However, Airbus hasn't really made any change to the design of this protection system, apparently judging that this was not a serious matter. According to the experts at Airbus, the way out from such an abnormality is simply to release control of the airplane, so that the automatic flight control system can take over and execute the go-around procedure up to around 3,000 to 4,000 feet above ground level.

Figure 1.1 (below) shows the final flight profile of the Nagoya accident. According to the National Transportation Safety Board (NTSB), similar to what was pointed out by the Japanese accident report, there wasn't any problem with the machine, and the automatic flight control system did everything according to the order, rejecting the override from manual control and executing go-around. Almost everyone agreed that this accident was caused by human factors.

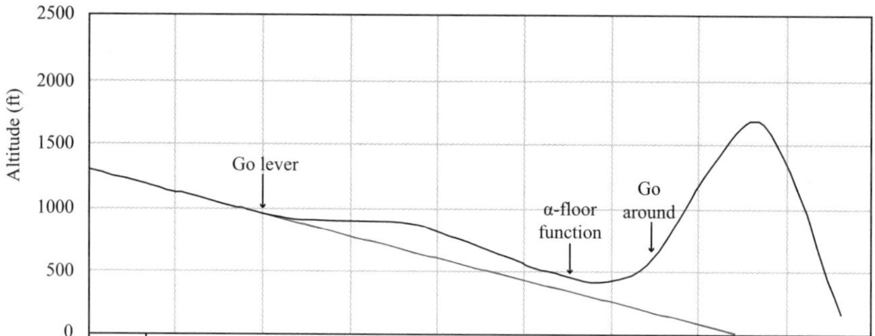

Time	13:50	14:05	14:20	14:35	14:50	15:05	15:20	15:35
Pitch		4.0	5.3	3.5	5.5	10.6	37.5	15.9
α		6.3	5.0	7.1	9.6	10.4	12.9	28.0
Elevator		0.3	3.5	8.5	9.9	14.1	14.1	0.6
THS		-5.3	-5.3	-12.3	-12.3	-12.3	-11.3	-7.2

Figure 1.1 The final flight profile of the Nagoya accident

For example, the F/O wrongly controlled the plane and the crew did not fully understand the functions of the automatic flight system. Hence, the corresponding solution is to emphasize the out-of-trim training to the pilots, just as suggested by the Transport Ministry of Japan.

But is this a sufficient explanation? Will similar incidents be prevented from happening again after enough training?

The primary goal of accident investigation is to find the real causes behind an accident, make recommendations, and ideally prevent similar events from happening again. However, after the Nagoya accident, in less than four years, an almost identical accident took place. The same type of aircraft crashed in almost the same manner in Da-Yuang, Taiwan. It forced us to think: have we indeed found the real cause of the Nagoya accident? And have the suggestions from the accident investigation report really mitigated the conflict between the pilots and the computer?

Every single human behavior has its origin. Some are after deliberate consideration belonging to the decision-making behavior. Some are from a cultural heritage that is shared within a community, some are from the sub-consciousness, perhaps even instinct. Behavior that appears on the surface to be the same might originate from completely different sources. If we insist on looking at only what it appears to be, and do not dig deep and get at the real cause behind such a behavior, how can we recommend an effective solution? Obviously, the principal beliefs widely accepted by the international aviation community might very likely bring about a self-restricting situation, causing the truth to be buried.

The cockpit voice recorder (CVR) is an essential resource for finding human errors. Further, the CVR can supply hints about the psychological conditions, even cultural characteristics, of the crew. The sequence of the Nagoya accident clearly shows the disastrous interaction between the pilots and the machine. How could the interaction become so fatal? What was in the crew's mind? What were the pilots thinking during the process? The main part of the conversation of the crew was in Chinese. Hence, it requires a thorough knowledge of Chinese language and culture, preferably by one sharing a similar cultural background, in order to have a thorough understanding of the real meaning of the conversation.

After the F/O inadvertently triggered the go lever, at the same moment the CAP reminded the F/O of the situation.

> 20:14:10 CAP: You, you triggered the go lever.
> 20:14:11 F/O: Yes, yes, yes. I touched a little.

At that time, the automatic flight control system started to execute the go-around procedure, and the plane began to level off. Yet the crew still wanted to land, so the F/O continuously pushed the control yoke, hoping to lower the nose. The angle of the elevator increased.

The CAP acknowledged the situation and commanded the F/O to disengage it.

> 20:14:12 CAP: Disengage it!

Immediately, the cockpit voice recorder recorded the sound of a pitch trim maneuver. Obviously the F/O was pushing the yoke and trimming the pitch, trying to disengage the automatic go-around which was already in progress, so that the plane could continue to land. However, the order that the computer received was to go around, so in order to raise the nose, the angle of the trimmable horizontal stabilizer started to rotate, countering the effect from the increase of the angle of the elevator. The elevator and the horizontal stabilizer started to diverge and their angles differed even more. The two control surfaces of the tail that should cooperate to fly the airplane actually started to go against each other.

Two contradictory orders continued to take effect on the horizontal stabilizer. After 20 seconds had passed, the CAP reminded the F/O that the plane was still in go-around mode, indicating that both of the crew recognized the go-around mode was still on.

> 20:14:30 CAP: You, you are using the go-around mode.

> 20:14:34 CAP: It's OK, disengage again slowly, with your hand on.

Then, the sound of pitch trim is heard continuously without a break five times, indicating that the F/O was still trying to disengage the autopilot and continue descending. Unfortunately, he was continuously trying the same ineffective maneuver. And on the contrary, it caused the out-of-trim situation, i.e., the conflict between the elevator and the horizontal stabilizer, to become even worse.

Since the CAP reminded the F/O that the go lever was triggered, 24 seconds had passed. During this time, the F/O had tried six times to disengage, yet in vain. Strangely, he never tried to speak out that he could not resolve the problem. Nor did he express that he did not know how to do it. According to the accident report, obviously the F/O didn't know the correct procedure, yet he refused to speak out or to ask, just continued trying the same ineffective maneuvers.

Several seconds later:

20:14:41 CAP:	Push more, push more, push more.
20:14:41 F/O:	Yes.
20:14:43 CAP:	Push down more!

Of course it was still ineffective:

20:14:45 CAP:	It is now still in go-around mode.
F/O:	Yes, sir.

The CAP reminded the F/O again that the plane was still in go-around mode. Several seconds later:

20:14:51 F/O: Sir, I still cannot push it down.

Strangely, another several seconds passed without any response from the CAP.

20:14:58 CAP: I, well, land mode?

Obviously, the CAP was asking the F/O if the go-around mode had been switched to the land mode, but the F/O remained silent. Another several seconds later:

20:15:01 CAP: It's OK. Do it slowly.

Yet right after he comforted the F/O, the CAP couldn't stand anymore and said:

20:15:03 CAP: OK, I have got it, I have got it, I have got it.

The CAP finally took over the control without waiting for the F/O to complete the disengagement of go-around. Yet the alpha-floor protection function had been activated. The throttle of the engine had gradually increased to the maximum. Since the action line of the thrust passes under the wing, right beneath the center of gravity of the plane, the increase of the thrust force resulted in a positive pitching movement. The nose was raised quickly, causing the pitch of the plane to further increase.

What happened next was really frightening:

20:15:08 CAP: What's the matter with this?

The F/O also panicked:

20:15:09 F/O: Disengage, dis …

The nose continued to pitch up, and the plane climbed further up.

20:15:11 CAP: Go level.
Damn it, how come like this?

Surprisingly, the CAP also did not know how to disengage the go-around! Meanwhile, the CVR recorded the sound of pitch trim twice. The CAP was trying exactly the same ineffective action that the F/O had already tried six times.

The pitch angle of the plane had been raised up to 21.5 degrees now, and was still increasing. The signs of an unusual and dangerous situation were very clear, so the F/O immediately called to the tower to prepare to go around.

20:15:14 F/O: Nagoya Tower, Dynasty going around.

Seconds later, the sound of flap movement was heard, the ground proximity warning system alarmed at the same time, and the angle of the airplane pitched up to 36.2 degrees. Facing such a frightening situation they had never experienced before, both of the crewmembers froze.

20:15:21 CAP: If this goes on, it will stall.
20:15:25 CAP: Finish.

In Chinese, this word means, "We're screwed." The stall warning went off, and the pitch angle reached an unbelievable 52.2 degrees.

20:15:26 F/O: Quick, push nose down.

In the cockpit, the atmosphere was extremely chaotic, all kinds of warnings and operational sounds clashed all together.

20:15:31 F/O: Set, set, push nose down.

The plane started to fall after it peaked to the highest.

20:15:34 CAP: It's OK, it's OK!! Don't, don't hurry. Don't hurry.
F/O: Power!!

The plane started to fall faster and faster … the crew was obviously terrified.

20:15:37 CAP: Ah …!!
F/O: Power, power, power!!

Until the end of recording, the CVR was full of frightened screaming. The CVR recorded deadly silence at 20:15:45. The runway of Nagoya Airport was already covered by a stretch of raging fire.

In earlier ages, there was not much computer technology used in airplanes. How the airplane flew was mainly affected by the aerodynamic features of the corresponding mechanical system itself. Even if the pilots were not fully familiar with the aerodynamic features of the plane, it would not have been a big problem because the airplane was not going to do anything by itself. As airplanes became more and more complex, together with the rapid development of computer technology, the modern large airliners quickly became automated. In plain language, the airplane itself started to do something solely on its own, autonomously. According to the flight control software, gradually it has its own power to make decisions, just like a real human being. So the aviation community called such an automated flight control system the "electronic crewmember," "electronic pilot," or "artificial pilot." Because this artificial pilot was created by the software engineer, it is the engineer who grants the artificial pilot the ability to think, to analyze, to judge and to make decision, to help the crew to complete the flying mission.

However, once the electronic pilot had the ability to make decisions, it started to develop its own viewpoint and accumulate its own power. Of course, its viewpoints are in fact the viewpoints of the flight control software design engineers, and its power was given by these engineers too. So, some say that the people who are really flying the airplane are not the pilots in the cockpit, but the software engineers. This is of course a little bit exaggerated; yet, it is definitely true in some aspects. At least, we can say so: the pilots and the software engineers must cooperate to fly the plane!

As airplanes become more automated, the electronic pilots will have opinions on more and more aspects of flight, increasing their power to fly the plane. Yet, most of the software engineers do not seem to know how to fly, nor do they have much flying experience. They just design the software according to the physical laws, flight dynamics and control theory. It sounds a little bit scary at first. Some engineers who cannot even fly a plane are actually designing the flying software. But on second thought, it isn't that scary. As long as the electronic pilots express their viewpoints concerning physical phenomena, there won't be a problem. Unfortunately, as the plane evolved to be more complicated, the software engineers began to equip the electronic pilot with more power to make decisions about non-physical things. Consequently, the flight control software extends its influence into the category originally belonging to humans. The psychological features, value system, and cultural traditions of the software engineers then started to intrude into the automatic flight control system. Thereafter, gradually, the electronic pilot, in a manner of speaking, bears white skin, golden hair, and blue eyes. It is a Caucasian.

The conflict between the autonomy of flight management systems and the human operators' authority has been recognized for many years (Sarter and Woods 1995 and Mumaw et al. 2001 are but two examples). Stated briefly, in an array of complex tasks, how does the designer and the operator allocate tasks between the human and the machine, and how are conflicts resolved? A recent study (Tessier and Dehais 2012) found that conflicts between human and machine can lead to reinforcement of behavior patterns and "attention tunneling" – in other words, a rigidity and narrowing of situational awareness that can reinforce a difficult situation. Such self-reinforcing behaviors were characterized in one study as a "death spiral" implicated in a controlled flight into terrain (CFIT) accident (Perez 2002).

Cultural differences add yet another potential source of conflict in this ongoing dilemma.

Finally, there came the day that the Chinese pilots and the Western electronic crewmember quarreled and fought over control of the plane, and both were opinionated and would not yield to each other …

How come this far from perfect Western electronic crewmember can cooperate happily and smoothly with other people? Even if there were some quarrels, they wouldn't be really serious or unsolvable. Yet in Nagoya, why would this Western electronic crewmember be so antagonistic with Chinese pilots?

Modern jumbo jets have all become highly automated, which includes very high levels of judgment and decision-making. How it judges or makes decisions was totally designed by the software engineers. Therefore, the psychological features, the values and the beliefs – and even the cultural traditions – of the engineers would penetrate into every corner of the flight control software. Given this, if we still stick to the superficial behavior of the crew when discussing the human factors without examining the influence of culture, it will surely not be realistic.

Let's take a look at a really simple example. After the F/O accidentally triggered the go lever, he tried many times to disengage go-around by pushing the control yoke and trimming the pitch. The CAP even took identical actions after he took over the plane. Why wouldn't they just give up after so many ineffective attempts? The real reason is that the pilots simply could not believe that such an action did not work. Why did these pilots have such a belief? The reason is very simple.

Before the Airbus started to manufacture airplanes, all jumbo jets of the Western world were made by Boeing. Pilots from all over the world flying the planes were familiar with Boeing's value system because it was the only system. The basic belief behind the design of Boeing airplanes about the final authority on the plane was that the final control of the airplane should be given to the hands of the pilots. No matter under what kind of urgent circumstances, pilots could take over the controls of the airplane whenever they wished, because Boeing thinks that it is the pilots who are flying the airplane. Pilots who are familiar with Boeing airplanes would naturally think so and act according to this belief, without noticing there could be another kind of value system.

By contrast, Airbus has a different perspective on this issue. With some oversimplification, Airbus thinks that ordinary people normally couldn't calm themselves and deal with problems wisely under very urgent situations. Further, compared with machines, humans are simply not as reliable as machines. To avoid such uncertainties from humans, Airbus has decided to yield the final authority over the airplanes to the electronic crewmember (the computer) under urgent circumstances, because Airbus thinks that it is the physical laws that are flying the airplanes. Of course it makes sense as well. It is, in fact, gravity, aerodynamics, and engine power that decide whether or not the plane can fly.

However, after the Nagoya accident, Airbus decided to yield the final authority over the plane under urgent situations back to the pilots. The reason for this was to prevent such a conflict between the pilots and the electronic crewmember from happening again. Airbus then changed the design of the flight control system. In other words, as long as the crewmembers pushed the control yoke down with more than 33 pounds of force, the autopilot would be overridden, and control over the airplane would be back in the hands of the crew.

There were cultural differences even between the two Western companies, not to mention the differences between the West and the East. However, compared with the cultural differences between East and West, the difference between Boeing and Airbus is negligible. To the Chinese pilots in Nagoya, they would not try to discriminate the difference of the controls on the planes of Boeing or Airbus instinctively, because they are all the same Western airplanes. Normally, the Chinese pilots and other company personnel would not expect there to be so much difference between Boeing and Airbus airplanes.

From the accident report, there are five points worth further consideration:

1. The F/O inadvertently triggered the go lever at 14:05. From 14:10 the CAP mentioned the go-around was on, and then at 14:30, 14:45 and 14:58 the CAP further reminded the F/O about the go-around mode. From 14:12, the CAP instructed the F/O to disengage the go-around completely five times, including another four at 14:23, 14:34, 14:41 and 14:43. However, the F/O did not respond to the CAP while he tried pushing the yoke and trimming the pitch six times. If the F/O responded and admitted that he did not know how to disengage the go-around mode after the first try, they would have had 40 more seconds to deal with the situation, even to call for help from the ground. The question is: why did the F/O not speak out that he did not know how to disengage the go-around mode after trying so many times?
2. The CAP instructed the F/O five times to disengage the go-around mode after the F/O inadvertently triggered the go lever. Though the F/O did not respond orally, he did try many times to disengage the go-around. The CAP was well aware of what the F/O was attempting. From giving the first instruction at 14:12 to the CAP actually taking over at 15:03, 51 seconds passed. Why did the CAP just watch and not try to help the F/O?

3. The F/O had tried many times to push the yoke and trim the pitch without success. After the CAP took over, he did exactly the same, pushing the yoke and trimming the pitch. Why would the CAP still repeat the same maneuver after the F/O had tried so many times? These behaviors, deeply rooted in the traditional Chinese authoritarianism, are not immediately evident within a Western mindset. However, they would not simply come down to earth by accident. Usually they are evoked by the environment having an appropriate combination of situations. Hence, the environment provided by the automated flight control system together with the mandatory standard operating procedure (SOP) should be rechecked if the cultural factors are to be taken into consideration.
4. When the elevator (controlled by the pilots) and the trimmable horizontal stabilizer (controlled by the computer) started to rotate in opposite directions, the result was surprisingly effective, despite being contradictory. The advanced automation expected to be the final authority and help the pilots fly the plane by itself, but it fell into an unexpected self-contradictory state. One can imagine how much confusion could be created in the minds of those two Chinese pilots.
5. Originally designed to protect the plane from stall, the alpha-floor function, after being activated, further corrupted the already dangerous situation. Facing such an unanticipated situation, an authority with apparently arbitrary will, how should the Chinese pilots react?

The automation designed by Westerners of course would not come down to earth by accident either. They would definitely have their roots in their culture. Such a flight environment created by Western cultures seems likewise to be non-comprehensible in the Chinese eyes.

Today, all jumbo jet airplanes are manufactured by Westerners, and every kind of operational procedure and requirement in the system is established by them. It can even be said that the whole flight environment is Western, from standard aviation English to Western-originated concepts such as crew resource management. Moreover, in every single cockpit, there is one Western artificial electronic crewmember ready to provide help in the Western way at any time they think necessary. What kind of conflict will be created when such a crewmember joins with two Chinese pilots in a narrow cockpit? To what extent can the conflicts be reduced through training with Western-designed standards? It would be difficult to expect Western engineers to understand Chinese culture and to avoid potential conflicts when designing the system. On the other hand, it would be just as hard to expect the Chinese to understand the Western culture and to act exactly the same as a Westerner would act when flying the plane.

Unfortunately, though it is really hard, it is unavoidable.

The Dragon in the Cockpit examines some of the fundamental differences between Chinese and Western cosmologies and value systems to suggest that some fundamental cultural collisions are inevitable when this Western technology is deployed within an Eastern context. Further technological refinements, as exemplified by flight management systems, have the potential for intensifying the cultural collisions rather than mitigating them. Evidence of this is found in the role of mode confusion and the flight management system in the Nagoya tragedy, as well as several others that we will examine in the next chapter. Cultures and cultural differences are sometimes presented as superficial decorations – tastes in food, mannerisms, speech accents – without an appreciation that these "habits of the heart" (Bellah 1985) are rooted in centuries, if not millennia, of learning and adaptation, and cannot be simply waved away. People, particularly highly trained professionals, *can* learn a new culture, just as they can learn a new language and become bilingual, but in stressful situations they typically revert to earlier, more ingrained patterns. This reversion to earlier cultural habits, Karl Weick has suggested, was implicated in the worst aviation disaster in history, the runway collision between a KLM and a Pan AM 747 in Tenerife on March 27, 1977 (Weick 1990).

Cultural differences determine the character of interpersonal relationships, modes of thinking, and responses to authority. When assumptions regarding these are hard-wired into technological systems, whether in the design of a flight management computer or the standards and procedures for accident investigation, misunderstandings are inevitable, although in most situations such misunderstandings can be translated and the parties can adapt and adjust. In situations where there are only a few seconds to spare – a miscommunication on short final, for example – the potential for damage is considerable. In more deliberative situations, such as an accident investigation, we have the opportunity to learn and adjust.

To present these points we first examine the cultural bases of some recent accidents in Chinese aviation. We show that responses to authority are often critical, and that this is a key difference between East and West. In Chapter 5 we examine the roots of Chinese assumptions about authority, both in history and in the figure of the dragon, an all-powerful, dangerous yet benevolent mythological figure. Research carried out by National Cheng Kung University measuring these features of authority is described in Chapter 6.

In Chapter 8 we present a new and fundamental concept for understanding Chinese interpersonal relations, the *guanxi gradient*. Guanxi gradient is a measure of the meaning of and investment in interpersonal relationships. Earlier research (Jing and Chang 2006) found a strong correlation between guanxi gradient and aviation safety, with cultures having a higher guanxi gradient having a worse flight safety record.

Are we therefore doomed to be prisoners of our own cultures? The struggles of these mindsets are described in Chapter 9, yet it is possible to synthesize their different approaches – the linear thought patterns based on an alphabetic language, and the figurative insights coming from an ideographic language like Chinese. If these cultures can learn from each other, the result will be a harmonious sky.

More than 200 years ago, in 1793, the British sent Lord Macartney to the Chinese court to demand an opening of China to British trade. At this time Great Britain was approaching the height of its imperial power, and the Qing dynasty was in a long, slow decline. These relative facts of imperial ascent and decay for the next century fostered shared assumptions of the superiority of Western technology.

Today the situation is reversed: an increasingly confident China is placing air transport and aircraft manufacture at the front of its efforts to assert industrial dominance, no longer relying on cheap manufacture for export. Western powers, by contrast, whether the European Union or the United States, are beset by paralysis and indecision. A harmonious sky could possibly guide the way for other paths of mutual learning and adaptation.

Chapter 2
Technology, Language, and Culture

Not more than four years after the Nagoya accident, another A300-600R plane crashed when landing. This time the accident was right in front of its own home.

On February 16, 1998, at 8:05pm (20:05), the CKS Airport control tower announced an emergency warning: China Airlines Flight 676 lost contact with the tower. The north and south stations of the fire department immediately sent out all the fire engines to look for the plane. A foreboding atmosphere filled the entire airport.

All fire engines and the corresponding personnel rushed out of the buildings but could not find any sign of the plane in the whole airport. Right at that moment, they heard the frightening voice from the tower: Flight 676 had crashed outside the airport! All the trucks quickly moved out of the airport. Five minutes later, the screaming voices of the firefighters came in through the radio:

> Fire! Fire everywhere! Full of wreckage and bodies! We need all support from every single unit.

Finally the tragedy was confirmed: Flight 676 had crashed in Da-Yuang, Taoyuan County, between #713 and #721, Section 2, International Rd., right next to the CKS Airport.

The day after the Da-Yuang event, some people from the local aviation community speculated about the cause of the accident, and pointed out that it seemed the altitude of the plane at the final phase of landing was too high, so that the plane couldn't land on the runway. Hence, the crew pulled up the plane and prepared to go around. Most likely, the conflict between pilots and the machine took place again. It may have happened like this: the automatic flight control could not be disengaged by manual control. Consequently, the plane couldn't ascend fast enough. In other words, what happened in Da-Yuang was quite similar to what had happened at Nagoya four years earlier.

Soon, as more and more information about the tragedy was released, the speculation became substantial. People started to discover that the Da-Yuang event not only resembled the Nagoya event, but was almost a direct copy of it. Were they really alike? Were they truly related? What was the reality behind the tragedy?

Two years later, the accident report by the aircraft accident investigation team of the Civil Aeronautics Administration (CAA 2000) was officially released. The report pointed out there was no mechanical failure with the plane and the weather conditions were good. Therefore, the Da-Yuang event was again attributed to human error.

The investigation team further concluded this accident was caused by the following:

1. During the entire descent and approach, the aircraft was higher than the normal descent profile and the pilot did not intercept the normal glide path.
2. During the go-around phase, the crew coordination between the CAP and the F/O was inadequate.
3. For 11 seconds during the go-around, the crew did not counteract the pitch-up tendency which was caused by the increase in thrust. The subsequent reaction of the crew was inadequate. As a consequence the aircraft's pitch increased until the aircraft stalled.

Again, it's go-around! Again, it is conflict between pilots and the computer! And again, it is pitch-up and stalls! It had already happened once in Nagoya. Just like reincarnation, the ghost came back again. What really happened this time, causing an almost identical tragedy?

According to the accident report and the flight data available, if we follow the theory of the accident chain of events from Boeing, the sequence of what happened in Da-Yuang was as follows:

1. The plane should have descended when the altitude was around 21,000 feet, causing the approach of the plane to be higher than the normal glide path.
2. When the plane passed the outer beacon, the altitude was 1,000 feet too high, while the plane still continued to descend.
3. The crew continued to lower the flaps and the gear, hoping to bleed off altitude rapidly, causing the speed of the plane to drop, yet the plane had to raise the angle of attack to maintain enough lift to stay flying.
4. At 20:04:59, the angle of attack crossed over the 11.5-degree threshold, the alpha-floor function was then activated, and the thrust of the plane began to increase to the maximum. The pitch started to increase accordingly.
5. The crew pushed the control yoke instinctively in order to lower the nose, yet because the force was more than 33 pounds, this caused the manual control to override the automatic flight control system. The autopilot was disengaged automatically.
6. The captain decided to go around right after that. In order to prevent the conflict between the pilots and the machine from happening again, the crew let go of the control yoke for the autopilot to execute the go-around. However, since the autopilot had been previously disengaged, no one was flying the plane – neither the pilots nor the computer. Hence, no movement of the yoke nor the elevator was detected for 11 seconds.
7. The pitch continued to increase until the plane stalled.

Figure 2.1 shows the final flight profile of the Da-Yuang accident.

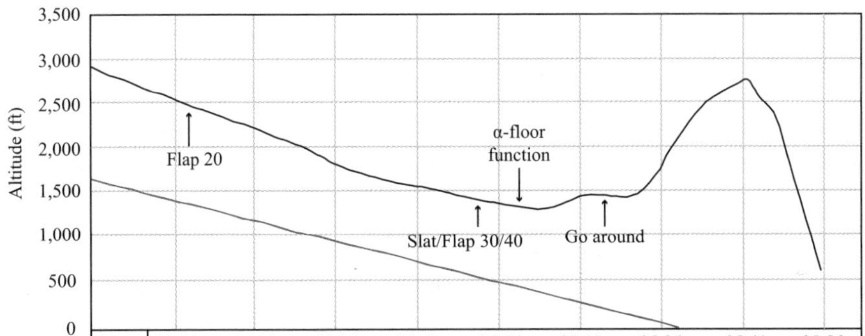

Time	03:56	04:11	04:26	04:41	04:56	05:11	05:26	05:41	05:56
Pitch	1.0	-1.0	-0.5	2.0	1.5	2.0	37.0	39.5	-18.0
α	7.5	5.0	5.0	7.0	10.0	7.5	18.5	40.0	11.0

Figure 2.1 The final flight profile of the Da-Yuang accident

Responsibility for investigation of the Da-Yuang event lies with the CAA. In the final part of the accident report, the CAA listed some suggestions for improvement to Airbus and the airline.

> To Airbus:
> Airbus should review the flight crew operation manual of the A300-600 to remind the flight crew the basic data of flight dynamics for a go-around, the procedures and crew task-sharing that permit the flight crew to perform an approach and a go-around safely.
>
> To the airline:
> 1. The airline re-examines and enhances flight crew training:
>   (a) Cockpit discipline and coordination;
>   (b) Clear defining of responsibilities between pilot-flying (PF) and pilot-not-flying (PNF);
>   (c) Positive acknowledgement and distinct call-out procedures;
>   (d) Adherence to standard operation procedures.
> 2. Airline conducts A300-600R simulator training in a more realistic manner for auto and manual go-around, normal and abnormal maneuvers, preventing and recovering from unusual attitudes.
> 3. Airline adopts and implements "airplane upset recovery training" to improve flight crew ability to confidently handle recovery from unusual attitudes.
> 4. Airline enhances the company safety culture by strengthening CRM training and improving communications between members of flight crew.
> 5. Airline adopts an ongoing education program to instill in crewmembers the conviction that human factors play a significant role in flight safety.

6. Airline must conscientiously and realistically carry out required training programs and inspection system.

In fact, regarding the efforts of China Airlines to implement the recommendations after the previous Nagoya event, there were two noteworthy paragraphs in the accident report.

1. After the Nagoya accident involving an A300-600R/B1816 on April 26, 1994, the Chairman, President, Director of Flight Operation Division, Director of Flight Safety Office and many others were replaced with a whole new management team.
2. Other than the restructuring of top management, the Airline signed a technical support agreement with Lufthansa Technik to provide technical support for the improvement of flight operation, engineering and maintenance, etc. The technical support was named as "Chi-Chiang (self-strengthening) Project" to do the following:
   - To enhance the comprehension of autopilot systems and its training,
   - To enhance the comprehension of aviation language and its training,
   - To enhance the crew's comprehension of aviation weather and the relationship of aircraft operations,
   - To enhance the crew's comprehension on aircraft performance,
   - To provide the training operation philosophy via cockpit procedure trainer and flight training device (FTD).

But the improvements were not significant.

Indeed, the improvement was not good enough. Otherwise, there would not have been a repeat of the Da-Yuang accident. Almost everyone believed that sufficient training was needed, yet it still could not be fully effective. The repeat of the Da-Yuang accident indicates that the real issue behind it isn't as simple as a lack of training. There must be something more profound and deep influencing the whole course of events.

According to the accident report, the first link in the chain of the Da-Yuang event is that the plane should have descended, yet it still flew level, causing the approach to be too high. The CVR clearly revealed this onset of the whole event.

19:35:46 F/O: I will talk to the cabin. There are seven minutes before descent.

As early as 7:35pm, Taipei local time, the F/O mentioned that the landing would begin in seven minutes, the CAP confirmed so.

19:35:47 CAP: OK. Thank you.

After nine minutes, the CAP ordered to land.

19:44:57 CAP:	Descend!
F/O:	Sir, my side …

It seemed that the F/O had something to say, yet without saying a word:

19:45:04 CAP:	Descend.
19:45:05 F/O:	Yes.

Therefore, the F/O started to contact the Taipei approach. It was already more than two minutes later than the scheduled time. If calculated by the actual speed, approximately 300 knots, if the plane kept level for two minutes, it would be 3,200 feet higher than the three-degree normal descent glide path. That is to say, even by very rough estimation, the plane was already 3,000 feet higher than the glide path when it started to descend. The CAP and F/O did not communicate with each other about this.

What followed in the CVR were only some normal communications between crewmembers and the tower.

19:47:36 CAP:	You see, if we don't descend, the visibility will be getting worse. Go back soon, it's important.
19:47:43 F/O:	Yes.

Not long after:

19:48:50 CAP:	Look at that side, see, there is the lightning quite strong.

Two minutes later:

19:51:14 CAP:	OK. Do you see how big the energy is? That is terrible!

Obviously, worrying about the weather, it is understandable that the CAP hoped to land quickly.

After the crewmember contacted the approach tower, the descent was further ongoing.

19:53:27 CAP:	Approaching FL150, we should report 2,000 feet to go, he hopes to let us keep descending.
19:53:30 F/O:	Yes.
19:53:31 CAP:	If we delay that we may have to change something.
19:53:33 F/O:	Yes.

It seemed that the F/O would say yes to everything from the CAP, without even giving it a thought, no matter what the content was.

Taipei local time 8pm, the F/O seemed to have discovered something and said with an inquiring voice:

20:00:35 F/O: Only 16 miles to the runway, currently the altitude is 7,000 feet.

If calculating the altitude according to the three-degree descent path, at 16 nautical miles to the runway the altitude should be 5,100 feet. In other words, the altitude of the plane at the current moment was 1,900 feet more than it should have been, and the actual difference might be bigger. Six seconds later, the CAP ordered to lower down the flap.

20:00:41 CAP: Flap 15.

The CAP did not respond directly to what the F/O asked earlier, yet ordered the F/O to lower down the flap. Lowering the flap would increase the drag of the airplane and thus deplete the speed and the lift accordingly. The consequence would be a larger sink rate of the plane, i.e., the plane would descend faster. So, it was obvious that the CAP heard the reminder from the F/O. After a few seconds' pondering, the CAP decided to increase the drag and decrease the speed to make the plane descend faster.

At this moment, the speed of the plane was lower than 200 knots. After one minute, the CAP gave the order to release the landing gear.

20:02:05 CAP: OK, gears down.
20:02:06 F/O: Gear down.

Obviously, the CAP still continued trying to increase the drag to let the plane descend faster. This time, the CAP used the landing gear to induce drag. But, according to the regulation, the landing gear can only be released after the flap is 20 degrees. Thereafter, the speed of the plane further decreased to 185 knots.

Meanwhile, Taipei approach tower discovered the altitude of the plane was too high, and immediately checked with the pilots.

20:02:18 Approach Tower: Dynasty 676, is it too high for you?

Very coincidentally, the pilots just switched the channel to the CKS tower. Hence, they did not receive this message. As the F/O kept contact with the tower, another minute had passed.

20:03:18 CAP: Still have two miles to go.
20:03:31 CAP: Oh, we are high. Go down further, it doesn't matter.
20:03:32 F/O: Yes.

Several seconds later, the CAP appeared to be a bit hesitant.

20:03:44 CAP: Oh, this won't work. Two times three is six, OK, two thousands. Two times six is 12. It should work. No, it's not OK.

The hesitation of the CAP was very obvious. Suddenly it's OK, but then it is not OK. Yet, after simple calculations, the CAP decided that it was not OK. Therefore, the CAP made up his mind to further increase the drag to increase the sink rate.

20:03:59 CAP: OK. Flap 20.
20:04:00 F/O: Flap 20.

The flap was further lowered down to 20 degrees, and the glide angle had been increased to 4.5 degrees, already 50% steeper than the normal glide path. Slowly, the speed of the plane further depleted, while the angle of attack thus gradually increased, 5, 6, 7, 8 …

The plane continued to descend, and the F/O also continued to keep contact with the tower about landing:

20:04:41 CAP: It's 1,000 feet higher.

The normal altitude should be 700 feet AGL at this moment. However, it was 1,700 feet. According to the regulation from the airline company, if the plane had not been stabilized at the altitude of 1,000 feet before landing, it should go around right away. Unfortunately, the CAP did not give an order to execute going around.

20:04:51 CAP: It is coming, 1,000 feet.

After considerable effort, there was still 1,000 feet higher at last. In the final stage, the CAP did not execute the go-around according to regulation; on the contrary, he further ordered to lower down the flap once more and continue to descend.

20:04:54 CAP: OK, 30/40.
20:04:55 F/O: 30/40.

The F/O continued to carry out the order from the CAP without any hesitation. He put the slat to 30 degrees, and the flap to 40 degrees, to deplete the speed even further in order to enlarge the sink rate one more step. Four seconds later, because the speed dropped further, the angle of attack increased further up to a certain threshold. Finally, the fatal alpha-floor protection function was activated. At this moment, the angle of attack reached the 11.5-degree limit, and the speed decreased to 167 knots; the plane had already entered into stall.

Once the alpha-floor function had been activated, the automated throttle will raise the thrust to the maximum, so that the plane could speed up to escape away from the danger of stall. And as the thrust was elevated to the maximum, the pitch of the plane was increased, so the nose started to become level and continued to be elevated, resulting in a larger pitch angle. At this time, it seemed that the crew noticed the plane had stopped descending, and the altitude of the plane was coming back, so the crew intuitively pushed down the control yoke, hoping to force the nose down to continue landing. Since the force applied on the yoke was greater than 33 pounds, the autopilot was disengaged and overridden by the manual control.

This 33 pounds threshold was from the correction after the lesson learned from the Nagoya accident. This correction was made to avoid the conflict between human pilots and the automatic flight control system over the final authority of the plane from ever happening again. The original design of Airbus was that the manual control could not override the autopilot during landing or going around when the altitude was below 400 feet. It was because of this design that the conflict in the Nagoya accident took place. Under pressure from different corners of the aviation community, Airbus revised this design and granted the final authority over the control to the crewmembers. As long as the crew pushed the control yoke with more than 33 pounds' force, the manual control could override the autopilot.

After this, the pitch-up tendency of the plane had been pressed down for a while. However, because the engine thrust had been elevated to the maximum, just after 10 seconds, the pitch of the plane built up again. The CAP noticed this situation and immediately ordered the F/O to go around.

20:05:13 CAP: Go lever, go around.
20:05:14 F/O: Go around, go lever.

Strangely, the F/O still adhered to the order from the CAP without any doubt, even though they had just completed the landing checklist in the last second. Moreover, they had not talked about under what condition they should go around, and how they should do so if they needed to. Obviously, the F/O had already gotten used to following any order from the CAP.

Then, the CAP further ordered to raise the landing gear:

20:05:18 CAP: Positive, gear up.
20:05:19 F/O: Gear down?
20:05:20 CAP: Gear up!
20:05:20 F/O: Gear up.

It was clear that the F/O had noticed and got very confused by the rapid change of the orders from the CAP. Even so, the F/O still did not have any suspicion or question anything. All the F/O did was to confirm the order. From a Western perspective, this might appear to be blind submission.

A series of operational sounds appeared in the cockpit. Right at that moment, the pitch of the plane had reached 18 degrees. The situation had already become very urgent: the plane was rushing up into the sky with maximum thrust and 18 degrees pitch.

Unfortunately, the most tragic coincidence took place at this crucial moment. In between 05:24 and 05:35, for a total of 11 seconds, there was no movement of the control yoke or the elevator detected. Besides, the crew did not adjust the pitch trim. The function of the pitch trim is to alter the angle of the trimmable horizontal stabilizer, so that it can move in accordance with the elevator to assist the pitch motion of the plane. The requirement of Airbus indicated that when the autopilot was disengaged, the crew had to adjust the trimmable horizontal stabilizer when using manual control to go around in order to equalize the pitch-up moment induced by the maximum thrust to keep the nose straight forward without being raised up.

However, there was no sign of movement from the trimmable horizontal stabilizer. The reason was because the autopilot had already been disengaged 20 seconds before without the pilots being aware of it. In order to avoid the conflict between human and machine from happening again as in Nagoya, the crew decided to let go of the yoke for the autopilot to execute the go-around. Such a coincidence resulted in a really bizarre, unimaginable, yet scary situation. In this crucial moment, no one was flying the plane. The autopilot was disengaged, and the crew let go of the yoke. Finally, the tragic combination produced an almost identical outcome as the Nagoya accident.

In addition to this, there was a very obvious feature during the whole course of the flight. The CAP was the one always talking and directing. The F/O almost had no opinion and only replied yes and acted immediately to everything the CAP said. Starting from 35:46 – when the CVR recorded "seven minutes before descent" from the F/O – until the end of recording, there were all of 30 minutes and 11 seconds. The F/O replied to the CAP with an unambiguous "yes" 17 times. The F/O had almost no questions about any of the CAP's decisions, nor did he appear to be monitoring the CAP, despite numerous errors and misjudgments by the CAP, and causing the mistakes to accumulate without being corrected.

But, it is not fair to say that the F/O had no idea about the mistakes made by the CAP. The F/O mentioned in 20:00:35 that there were only 16 miles to the runway when the altitude was 7,000 feet. This indicated that the F/O did realize there was something wrong with the altitude. In the entire course of flight, given so many violations by the CAP, there must have been much suspicion in the mind of the F/O as a trained professional pilot. The real problem is: why didn't the F/O speak out and express his concern? Why didn't the F/O question the mistakes of the CAP? It's obvious that the CAP repeatedly violated standard operating procedures. The CAP did not even follow the rule when going around, yet still insisted to land at the final stage. Why didn't the F/O challenge the CAP? Can this be interpreted as blind submission? What if the F/O thought in the same way as the CAP?

Obviously, it is, again, traditional Chinese authoritarianism that brought all of these contending forces together. Two Chinese pilots sitting in a Western cockpit would not make them Westerners simply with the standard operating procedures (SOPs). They are still 100% Chinese.

In addition to traditional Chinese authoritarianism, there is one, even more profound, reason that caused the distortion of the SOPs: the written language.

The basic mode of thinking of Western people can be described as "sequential" or "linear," nurtured by the alphabetical language. There is always a fixed sequence to write and pronounce every single English word. And it is completely different from the Chinese. The written characters of the Chinese language are invariably understood ideographically because the sounds of them simply do not carry enough information. As long as the graphic image is completed, it doesn't matter what kind of sequence is used to write and what kind of sound to pronounce. It would be totally acceptable as a communication tool. To all the Chinese, a very strict and fixed sequence of talking and writing is hardly imaginable. Describing the profound impact brought by the Chinese characters, the basic Chinese mode of thinking is therefore called "ideographic," without any emphasis on the sequence nor the sound. Hence, it is very easy for people using an alphabetical language to follow standard operating procedures, while it is not the case for the Chinese, because the sequence is not in the instinct of the Chinese. Procedural, linear thinking fits better with an alphabetic, linear language. Moreover, the Chinese will have to fight with the pressure from culture so as to strictly follow the standard operating procedures.

The written languages not only affect how the Chinese behave, but also influence the Western culture, hence, the entire flight environment including flight safety.

Based on the alphabetical type of written languages, Westerners have a basic tendency to explain flight safety from a sequential point of view. Therefore, almost all the flight safety theories they propose emphasize sequence. The accident chain theory of Boeing is a typical example. Of course there are also many other similar theories. The accident chain theory arranges the occurrence of the accidents as a causal chain of separate events. According to this, it is very natural to infer that if one of the links in the chain is removed, the chain will be broken and the causal relation will be interrupted so that the accident can be stopped. After the Nagoya event, theoretically, when the link "human machine conflict" was removed, we should be able to break the chain and prevent similar accidents from happening again. Unfortunately, it was not as simple as that.

After the Nagoya event, the automatic flight control system of the plane had been revised to avoid conflict between human and machine. Yet in the Da-Yuang event, it was exactly because of this reason that the plane became uncontrollable at a crucial moment, causing an almost identical tragedy. Therefore, this sequential type of flight safety theory bears obvious flaws. Since this type of theory can only supply sequences, which are just lines, it is then very difficult to reveal the whole picture of the accident by simply drawing lines only. In typical aircraft accidents, there are usually many events that occur simultaneously. Which route will materialize and become the real chain of event depends on so many factors.

If one of the routes is broken, there are still many other routes by which the accident could occur. Sometimes, through bypass, the chain will go back to the original route and link to a similar outcome. After all, the accident will find its own way.

January 30, 1995, it's the last day before the Chinese New Year. The joyful atmosphere of the New Year had already been everywhere for a while. Planes, trains, and buses were crowded full of passengers anxiously going home for New Year's Eve dinner with family members. The people living on the small islands outside Taiwan were even more eager to find transport because they did not have many choices. Most had to rely on airplanes.

In order to get everyone back home in time, the CAA had already worked out a transport plan for 600,000 passengers, asking every single airline company, especially those flying to the isolated islands, to let everyone get back home before New Year's Eve for their family reunions. In coordination with the CAA plan, all the airlines dedicated every resource to help transport the passengers.

A twin-engine, turboprop passenger plane built by ATR in France was making a flight specially arranged by TransAsia Airways for the New Year. At 6:00pm, it was scheduled to fly from Taipei to Magong, Penghu Island. Seventy-four passengers filled the narrow cabin. It was the tenth flight to Magong that day. Everything was going very well after departing from Taipei.

At 6:50pm, the plane landed at the airport of Magong. After a short while, the crewmembers and the flight attendants walked back to the plane joyfully. They had just received a call from the company saying that the last flight from Taipei to Magong had been cancelled because all the passengers had already been transported back. This meant that the crew could go back home to have dinner with their families after the last flight from Magong to Taipei. A joyful atmosphere filled the plane right away.

At 7:16pm, Flight 510A departed from Magong Airport without any passengers on board, scheduled to land in Taipei at 8:20pm.

Sixteen minutes later the crew called Taipei approach, said the plane was 16 n.m. from Miaoli Houlong, and would descend and maintain an altitude of 9,000 feet. The approach tower directed the crew to pass Houlong and descend to 4,000 feet, and prepare to land on runway 10 of Taipei Songshan Airport. When the plane passed through Houlong, the crew asked to fly directly to Linkou, and proceed with visual approach. The approach tower agreed and directed the plane to lower the altitude to 4,000 feet.

At 7:40pm, the approach tower called again and directed the plane to descend to 3,000 feet. The crewmembers received the message and repeated it back. Two minutes later the crew said they already had the Songshan Airport in sight. The approach tower directed the plane to approach visually, and land on runway 10. The pilots confirmed the permission and repeated. One minute later, the pilot called the airport indicating they were still 15 n.m. to the airport. Until this moment, everything still seemed normal. However, after 7:44pm, the tower contacted the flight repeatedly but without any response. At 7:53pm, the flight disappeared from the radar screen.

Two years later, the CAA published the accident report (Civil Aeronautics Administration 1997). In the report, there were no primary causes revealed. Only the conclusions of the investigation were listed. Among them, several main points were given as below.

1. There was one flight attendant sitting in the cockpit.
2. The weather in the Songshan Airport met the requirements for visual approach.
3. The crew did not follow the standard procedures for calling.
4. The CAP did not fully understand the work and responsibility between the pilot-flying (PF) and pilot-not-flying (PNF).
5. The crew were not very familiar with the topography around Songshan Airport.
6. The CAP discovered that the ILS (instrument landing system) signals were not correct and yelled "No", "No", "Don't deviate".
7. The autopilot was disengaged two seconds before the crash, and in the final second, the digital flight data recorder (DFDR) recorded the G force of the plane increased to 1.8 as the nose was pulled up.

According to these conclusions, the CAA proposed some recommendations to the airline for improvements.

1. Reinforce discipline in the cockpit.
2. The sharing of the workload and responsibilities of PF and PNF should be very clear.
3. The crew should always comply with the SOP when operating the plane.
4. The training program should be re-examined. It is necessary to encourage training courses about the equipment and functions of the ILS system.
5. Review the SOP and start the training of the operation procedures of cross-check for different phases of instrument landing and other navigation aids.

According to the conclusions and recommendations in the report, the real cause of the accident was very clear. In the report, the first point mentioned was cockpit discipline of the crew, indicating that the violation of the rules was crucial to the accident. Why was there one flight attendant sitting in the cockpit? And why didn't the crew make the call according to SOP? In the report, another point was raised: the crewmembers were not very familiar with the area around Songshan Airport. However, this might not be true since the CAP had flown the route seven times and the F/O six times in this single day. It didn't make sense to say they were not familiar with Songshan Airport. The most probable explanation would be the carelessness of the crewmembers.

But the crewmembers weren't so careless so that they were totally unaware of the situation. The report showed that the CAP did realize there was something wrong with the instrument landing and yelled out "No", "No", "Don't deviate."

The real question was: why wasn't it corrected? Until the last second, the G force was increased up to 1.8, and the nose was already pulled up. It indicated that in the last second the crewmember found that they were flying too low and tried to pull up. There were only 25 feet for them to cross over the top of the mountain. They did not have time to prevent the tragedy.

Why and how did this happen? From the report of the CAA, it can be found that the CAA again attributed the cause to lack of training. Therefore, reinforcing crew training can prevent such tragedies from happening again. However, after every single plane crash, accident reports always suggest to reinforce training. "Reinforcing training" is almost a cliché in flight safety of Taiwan. Yet, plane crashes continue to happen. Why is that? Is it because the training is still not enough? Or, is it because that reinforcing training is very hard? Is it possible that training is not the real cause? Or were there some more profound explanations other than training? What is the truth?

From the CAA report, we can infer some cultural factors in the accident. Although these inferences are not part of the original report, they are consistent with the events of that night and with what we know about Chinese culture. In the accident report from the CAA, the conversation between crewmembers, the radio communication between crewmembers and traffic controllers, together with all kinds of sounds, were disclosed. From 13:23 after 7:00pm when the flight was approved by Magong for take-off, until 43:57, the total length is 30 minutes and 34 seconds. From the moment when the plane was still on the ground, there was the voice of a flight attendant (F/A) in the cockpit.

13:34 CAP:	Come on! Show us what you can do.
13:37 F/A:	(laughing)

It's obvious that the CAP was telling the F/O to practice taking off. At this time, the plane was still speeding up on the runway. Judging from the laugh of the flight attendant, it seemed that they were very familiar, so that they could stand the laughing during take-off. In fact, such relaxed conversations occupied almost every phase of this journey.

Yet, when there was one extra person in the very narrow cockpit, carrying on a lively conversation, it would definitely be very hard to preclude the possible consequences.

30:26 CAP:	Papa.
30:31 F/A:	What is Papa?
Controller:	TransAsia Airways 510A say speed?

Right at the moment when the flight attendant was asking the CAP what does Papa mean, the controller in Taipei approach tower inquired about the speed. What followed was the sound of channel switch, which might affect the crewmembers' ability to receive the messages from the controller. But, the crewmembers should have heard that since the messages were recorded by the cockpit voice recorder (CVR). Yet at this time, the CAP only replied to the flight attendant, not the traffic controller.

The conversation between the flight attendant and the CAP continued.

42:26 F/A:	Sir, approximately at what height when descending, can we lower the gear?

Before the CAP had time to reply, there came in the question from the controller again. It was still about the speed.

42:31 Controller:	510A say speed?

Yet the CAP chose to reply to the flight attendant first. Meanwhile, the controller asked again.

42:34 CAP:	Yeah, depends on the speed. Look at the speed while approaching certain distance.
Controller:	510A say speed?

Several seconds later, the CAP then replied to the controller.

42:40 CAP:	245.

The altitude of the plane was already as low as 2,400 feet. The traffic control was switched from Taipei approach to Songshan airport tower, and the CAP then contacted the Airport tower.

42:54 CAP:	Songshan Tower, 510A, 15 miles visual approach.

The Songshan tower reminded the crew of the wind direction and speed of the airport. The CAP repeated and reminded the F/O:

43:20 CAP:	Hurry up! Don't delay! Flight 522 is behind us preparing ILS.

CAP reminded the F/O that there was another plane waiting for landing. Yet, the F/O did not strictly comply with the SOP from the very beginning. Since the CAP handed the control to the F/O, he did not complete the checklists such as after take-off, cruising and descent according to the regulations. And the CAP, who supposedly should be monitoring, did not raise any correction either.

Seconds later:

43:37 F/O:	Capture …

The F/O called out "capture" unexpectedly yet the exact meaning was not very clear. It is presumed that the F/O was intercepting the glide path. Because at this moment, the recorded data revealed that the deviations of the localizer and glide slope were diminished quickly, although still far from correct.

After only few seconds:

43:40 CAP:	... No! No!
43:44 CAP:	Sir, don't deviate!

The CAP called out "No" twice and "Don't deviate!" This might be because he discovered the ILS message showing that the glide path of the plane was not correct. Right after this, there was five seconds' continuous warning of the altitude, which might be because the height of the plane was already too low, and yet the landing gear was still not lowered down. However, the response of the F/O was not clear, although the altitude of the plane did change. It seemed that he was making corrections. Unfortunately, it appeared that the F/O was still executing the interception.

43:56 F/O:	Capture ...

There was only one second after the F/O called "Capture" again, and then the recording was ended. With the end of recording, however, the waiting for their family members to join the New Year's Eve dinner for those four families would never end. What they got were urgent calls from the company and bad news on TV. New Year's Eve was frozen forever for them.

According to the investigation report by the CAA, with the DFDR and the CVR, a rough picture of this accident could be portrayed. The major cause of this accident was, according to the CAA, a failure of cockpit discipline. The discipline issue was a fact, yet what reason caused it? Why wasn't the discipline abided by? Why was it so serious as to cause the accident?

Collecting all of this information, there are at least several points listed below worth further consideration:

1. Why was a flight attendant sitting in the cockpit?
2. From 13:37 after 7:00pm, when the flight attendant spoke for the first time, until 42:26 when the flight attendant asked the last question – in total 28 minutes 49 seconds – the flight attendant had spoken 28 times. The CAP replied 21 times, and the F/O three times. Why was the conversation so frequent?
3. At 30:31 and 42:31, the flight attendant interfered with the communication between the crew and the controller. The crewmembers did not stop the flight attendant from interfering; on the contrary, they replied to her question first.

4. The CAP handed control over the plane to the F/O, yet when the F/O did not comply with the SOP in many ways, why didn't the CAP correct the F/O?
5. At 43:40, after the CAP yelled out "No, No", the F/O did not yield the necessary corrections to intercept the glide path. The plane still continued on the wrong path. Why didn't the CAP take over the control or stop the F/O since he already knew it was wrong?
6. At 43:44, when the CAP called again "Sir, don't deviate", the F/O still did not respond to the point while doing an unsuitable correction. Why didn't the CAP do anything? Why didn't he take over the control?
7. The CAP called the F/O "Sir". According to the investigation, the CAP was in the class of '43 of the R.O.C. Air Force Academy while the F/O was in the class of '41. The F/O was senior to the CAP in the R.O.C. Air Force Academy by two years.

The above points might seem a bit obscure, but anyone familiar with Chinese culture could find some hints from the traditional value system in Chinese culture. In other words, we can always perceive the connections between the above phenomena and our culture.

Why was there one flight attendant sitting in the cockpit? On that journey, everyone could be familiar workmates for many years and even good friends. Besides, it was New Year's Eve, a cheerful atmosphere was everywhere, even up in the sky. Further, there were no passengers in the cabin, which means no passengers for the flight attendants to serve. Moreover, they had completed their assignments, preparing to fly home. How could any Chinese be so strict as to push a good friend out of the cockpit in such circumstances just because of cold regulations? In Chinese culture, preference given to an interpersonal relationship, or "relationship favor," is a very important and undeniable factor in any interpersonal interaction. Sometimes it is a kind of resource for social exchange, and sometimes it is a pressure hard to escape. And it's common for Chinese in many aspects to distort the rules, regulations and even laws because of the exchange of a little bit of relationship favor.

Why didn't the crewmembers stop the flight attendant when she interfered with their communication to the controller? In Chinese culture, who would treat a good friend like this given such a relaxed atmosphere? Again, relationship favor played a very important role in the behavior of the crew. As for the reason why the crewmember replied to the flight attendant first, in Chinese culture, the relationship favor is hierarchical. If there are two people, one is a good friend, the other is a stranger, the two stand at quite different psychological distances in our mind. Every Chinese person would be inclined by instinct to attend to a friend first, not the stranger.

Since the CAP handed the control to the F/O, the F/O was the pilot in charge of flying, and the CAP, the pilot-not-flying, in charge of communication and monitoring. The F/O did not comply with the SOPs in many aspects – why didn't the monitoring pilot correct him? It might also be because of relationship favor, but we will never know what the truth is. Yet there is one thing we are certain of. If the CAP and the F/O did not get along well with each other, or if there was resentment between them, the CAP would definitely correct the F/O and even report to their superior when the F/O did not comply with SOPs.

At the final phase of landing, when the plane got closer to the ground, in a rather dangerous situation, the F/O made two mistakes consecutively. Why did the CAP just remind the F/O orally? Why didn't the CAP stop the F/O or even take over the plane? The CAP is not only the pilot monitoring but also superior to the F/O. There should not be any difficulty in him executing his monitoring duty. When the CAP called the F/O "Sir", a word in Chinese conveying a hierarchical sense and carrying the meaning of power, the reason is apparent. In fact, the CAP was not superior to the F/O, but the F/O was two years' senior to the CAP when in the Air Force Academy.

Superiority creates a power difference in any organized task. But, the key is where to locate the superiority. In Chinese society, the superiority does not simply come from official position or status. Sometimes, relationship does provide some alternative form of superiority.

It would be very hard to point out mistakes from a "Sir" in Chinese culture. It would be even harder to produce the corresponding actions. When facing mistakes from a superior, the pressure originated from Chinese hierarchy is established naturally. How big would the pressure be? It depends. However, the important thing to note is that some hierarchical pressure from relationship is hard to see. Even when the junior had become the CAP, the pressure was still big enough to prevent him from carrying out the action to stop his one-time superior although he was just his F/O now.

In modern, Western cultures there is a fundamental belief that all men and women are created equal. There is an individualistic ideology, from the influence of Christianity. Conflict is thought to be a good thing and should not be eliminated. Moreover, it should even be designed into the institution as the motivation of progress. Models of task-sharing in the cockpit between the CAP and the F/O were designed from the viewpoint of this ideology, and the belief system behind it. When the system asks the pilots to correct the mistakes of the other one in the cockpit according to the regulation, neither the one who raises the issue nor the one who receives the correction would feel uncomfortable. They wouldn't feel that they are losing face or get offended, because both of the two parties, together with the system designer, possess the same value system. Hence, when the system asks the CAP to be in full charge, to take over control of the plane from the F/O when necessary, normally, the system can be expected to operate smoothly.

However, Chinese culture is of an ethical type with situationalism. It always emphasizes harmony for every different kind of situation so that everyone should perform different roles under different circumstances. No one wants to ruin the harmony no matter how superficial it is. From the long history of China, only a harmonious society can unite people to struggle with the harsh nature of progress and development. When two Chinese who subconsciously believe that harmony is the way everything should be – and should also be maintained even in the narrow cockpit – face an environment created by people with completely different beliefs, how much pressure will they face when the SOPs conflict with supreme harmony? When the regulations ask them to correct the mistakes of good friends or even seniors at any time, or even to take over their jobs regardless of the relationship, favor and friendship, what would be the obstacle they need to overcome? How big is the chance of success?

Chapter 3
Fundamental Features of Taiwanese Accidents

Aviation technology was invented in the West, and the air transport system was built by Westerners as well. Contributions from other cultures are almost nonexistent. In the development of aviation technology, it was Westerners who initially confronted and solved some of the most basic problems, problems of entry into unknown territory, whether of flying at higher altitudes or placing new demands on crew performance. For the Chinese, aviation technology comes from abroad, and the corresponding problems are of course coming from how to use the technology. Therefore, the nature of Taiwanese aircraft accidents is then completely different from that of the West.

3.1 Causes from the Whole Company

The earliest air transport in Taiwan was initiated during the Chinese Civil War. Along with the fade-away of war, the development of air transportation gradually changed from being controlled by politics to technology. Sixty years later, repeated plane crashes sent Chinese aviators some very subtle messages: it seems that we do not really know how to handle Western aviation technology. Here, we will use the most representative cases of Sanyi and Hsinchu to explore the different traits of Taiwanese aircraft accidents.

On August 22, 1981, a Boeing 737 took off from Taipei at 9:55am, heading to Kaoshiung. There were 104 passengers on the plane with six crewmembers and flight attendants. About 10 minutes after take-off, at the altitude of 22,000 feet above Sanyi, Miaoli, the plane disintegrated in the air and exploded. Of the passengers and crew, all 110 on the plane died.

At 10:08am, the radar screen of the air traffic control showed the target light-spot split into two; in a minute, the target disappeared. Around 11:00am, the ground report came in, the plane had crashed near Sanyi with no survivors. According to the rescue personnel, the wreckage and dead bodies at the scene were scattered along an 8km long area, the nose, the door, the fuselage, the wing and the tail were torn apart and found in different places. Combining the radar and how the wreckage spread, it seemed the plane disintegrated in the air.

After a prolonged investigation, the CAA released the report. According to the wreckage found at the scene, the Sanyi accident was caused from a fracture in the bottom of the fuselage resulting in leakage of cabin pressure.

Because there was still a huge pressure difference inside and outside the cabin, the plane's structure experienced a bending failure, then broke all the way down. The fracture in the bottom of the fuselage was caused by widespread corrosion. Parts of it were even perforated, delaminated and peeled. Combined with the repeated pressurization of the cabin, fatigue cracks formed in different places. As time passed, with their growth, mutual coalescence, and development, combined to form structural damage. The plane broke apart. Therefore, according to the CAA report, the primary cause of the Sanyi accident was the metal fatigue fracture developed from corrosion.

The original assignment of this plane was to fly to Magong, Penghu Islands at 7:20am. Ten minutes after take-off, the crew identified some problems with the cabin pressure, then returned back to Songshan Airport, Taipei. Passengers on board were changed to another plane to continue the journey, and this plane stayed in the airport for inspection. After maintenance was completed at 9:20am, the plane then took on another assignment to Kaoshiung at 9:30am. The airline company which owns this plane operated four flights from Taipei to Magong and 10 flights from Kaoshiung to Magong every day. According to the report, this plane frequently transported seafood from Penghu back to Taiwan: the salty sea water dripped continuously on the cabin floor and caused the metal structure to corrode.

The CAA pointed out in the report that the maintenance personnel in charge did not carry out the regular anti-rust spray to the floor of the cabin according to the original requirements from the manufacturer. Instead, they used an alternative method – a bottom layer paint which did not have enough anti-rust performance to coat the cabin bottom – and so the metal structure kept on being corroded. Besides, the regular inspection of the fuselage structure failed to discover the corrosion, and hence failed to prevent the growth of fatigue damage. Therefore, it is obvious this accident was caused by maintenance human error. Consequently, the CAA proposed some recommendations to the airline as follows: maintenance should carry out the anti-rust task regularly as advised by the original manufacturer, and execute the cabin structure inspection from inside out regularly, using x-ray inspection as the main tool to find any cracks in advance.

However, there was one really striking sentence in the report released by the CAA: "within three months before the accident, the pressurization system failed consecutively for 21 times." This message was really astonishing. How could it possibly happen that the pressurization system failed 21 times within three months? It clearly indicates that the problem had never been solved. How could the airline management have no concern at all? How could they keep letting the plane leave the ground and carry passengers? Hadn't this chain of severe warnings raised any attention at all? Or is it that they really do not understand how severe the situation was?

The vice president of the airline said:

> Pressurization of cabin solely would not cause the explosion. Because this kind of pressure adjustment is designed for the human body, the goal is to let our body

> feel just like it is still on the ground. Therefore, even though the pressurization system did fail, as long as the plane descends to 8000 feet height, people can still adjust to it.

From what the vice president said, it seems clear that he did not have any understanding about metal fatigue. He treats the pressurization system like the thermostat of an air-conditioner. The main purpose is just to adjust the temperature: if it gets too cold, turn the air-conditioner off. So, when the cabin pressure isn't at its right level, just lower the height. He had no clue that every time the airplane was pressurized, the plane's structure would be stressed and then pulled once, and the flaws within the metal would increase.

The vice president might have been highly educated as a manager, but without any engineering background. It is possible that he had no clue about metal fatigue. Yet, on that day, the captain, who originally flew the assignment to Penghu, who too found out the problem with the pressurization system and returned to Songshan Airport, surprisingly said: "It shouldn't have caused such a major disaster if it is just the breakdown of pressurization system." A person who was already a captain with a flying background and lots of experience, and returned the plane back to the airport because of the pressure problem, seemed to not know how severe the problem could be with the metal fatigue, especially after so many times of consecutive system breakdown. It is really hard to imagine.

The CAA report clearly stated that the cause of the Sanyi accident was the human error of the maintenance personnel, so the CAA recommendations pointed to them. However, the fact that the plane was still in operation after 21 consecutive system breakdowns within three months is definitely true. It says that even the maintenance personnel were not cautious enough to understand the severity of this issue. That is to say, from top to bottom, from the vice president to the captain, and to the maintenance personnel, no one seemed to know how severe a problem metal fatigue could cause. If there was anyone among them who knew, they would have known that the cabin pressure is the source of the metal fatigue, and the metal fatigue could pull the entire structure apart. Had they understood this – facing 21 consecutive breakdowns of the cabin pressure – they would have been shocked by the potential dangers and would never have dared to let the plane fly.

Therefore, the real cause of this accident should be: when we brought in this huge complicated machine, we didn't really acquire the necessary knowledge we should have had before we started operating it. Thus, no one in the company really knows the seriousness of metal fatigue. So, the real cause of the Sanyi accident should be the lack of knowledge, not only the negligence or mistake of the maintenance personnel. Because if the cause is negligence or somebody's mistakes, it implies that they knew it, they just missed or forgot it. Unfortunately, in the Western aircraft accident classification systems, there is no such cause as "lack of knowledge." Therefore, in interpreting the human error according to the Western classification system, we need to be extremely careful, or else it is very likely that we will misunderstand the truth and be misled.

The knowledge that was lacking is what the philosopher Michael Polanyi characterized as "tacit knowledge" (Polanyi 1967). In contrast to explicit knowledge, tacit knowledge is not codified or captured in manuals or specifications, but rather acquired through experience. The Western aviation industry initially learned about structural fatigue not from consulting tables on the properties of aluminum spars, but from the experience of the first jetliner, the deHavilland Comet, which after being introduced into flight in 1949, suffered three unfortunate mid-air disintegrations before engineers figured out that the design of its square windows exacerbated the already unfamiliar stresses of high-altitude flight. Once it was learned – through experience – that metal fatigue due to cabin pressure differentials were an issue to be engineered, subsequent jetliners did not have this problem. But the problems of metal fatigue, while now codified in design manuals, are also part of the tacit knowledge of aerospace engineers.

The occurrence of the Sanyi accident indicated that we very often operate alien technologies under ignorance, since we usually have no idea about the mindset and the corresponding design logic behind those technologies. Some people may argue: weren't we provided training and handbooks, manuals, when we buy these Western technology systems? Obviously, we were. It's just that we might not really know where to look when we operate them. And we might not have the same operating habits as the Westerners do and, most likely, we will have a totally different mindset. Therefore, we are often faced with unexpected consequences coming from unanticipated human machine interactions unforeseeable by the original designers.

We will often discover that there is something we need to adjust internally within our system, even in our mindset, after we have paid a very high, sometimes tragic price. Then we would be able to operate the Western aviation technology. Without a doubt, it's an issue of evolution and adaptation.

3.2 Causes from Society

When an aircraft designed by culture A is crashed by culture B, it does not always come from lack of knowledge. Many times it originates from some more profound reasons. In other words, even when our airline companies have enough explicit knowledge, and know how to operate these technologies, if our society does not possess the right attitude about these Western technologies, we won't know what latent hazards still exist. The Hsinchu accident is just such an example.

It was March 18, 1998. A Saab-340 took off at 19:29, flying from Hsinchu back to Kaoshiung. There were nine passengers in the plane and four crewmembers. After departure, the co-pilot contacted the Taipei approach control reporting the altitude of the plane at 1,000 feet, and would continue to climb to 3,000. The Taipei approach control instructed to keep 3,000 feet and radar heading 260 degrees. At 19:31, the Taipei approach control directed the plane to turn left to 230 degrees. The co-pilot read back the instruction. Yet at 19:31:38, when the approach control asked to confirm the heading, there was no response. The Taipei approach tower tried calling the plane several times, all in vain. Finally at 19:32, it was confirmed that the plane had disappeared from the radar screen. The plane had already crashed into the sea with no survivors.

Three years later, the CAA released an investigation report (CAA 2001) that listed six possible causes of the accident:

1. The flight crew could not maintain situation awareness, resulting in a loss of control of the plane.
2. The right main bus broke down, putting the navigation and instrumentation of the right side out of order.
3. The flight crew did not comply with the requirements from the minimum equipment list.
4. The flight crew lost the visual reference from outside at night and under instrument meteorological conditions.
5. The captain flew under fatigue and space disorientation.
6. The flight crew did not comply with the standard operating procedures.

In the meantime, the CAA made four proposals:

1. The airline should enforce the discipline and requirements of safety regulations.
2. The airline should enforce the daily check of flight operations and establish the flight operation quality assurance system.
3. Reinforce the crew resource management training, abnormal condition and upset recovery training and system training.
4. Reinforce the monitoring of the regulatory body to the airlines.

From the possible causes listed in the investigation report by the CAA, the primary one of this accident was again the human error of flight crew. From suggestions proposed by the CAA, we can understand further that human error here meant the lack of compliance with the minimum equipments list (MEL) and with standard operating procedures. Therefore, the training of flight crew should be reinforced. Of course, the investigation report was firmly founded, based on the actual facts, yet the final judgment about the cause of the accident followed the typical classification from the West.

According to Western thinking, the accident causes can only be based on actual facts, and those actual facts were aligned as the events in the causal chain of the accident. For human error, the actual facts meant the actual observable human behaviors. Westerners definitely realize what causes are behind their behaviors, since they have the same common habits, values and beliefs which are the origin of human behaviors. Hence, the recommendations derived from those behavioral classification systems should normally be enough to help prevent similar behaviors from happening again because both the classification and the behaviors are based on the same value system. However, when the attribution of behavior comes across a cultural boundary, inference should be made carefully: the same behavior could have different causes behind it. Consequently, the classification system, based on the beliefs of Western civilization, are often insufficient, and sometimes even misleading when applied to different cultures having different value systems and beliefs – for example, Taiwan. This mis-application will cause these recommendations to lack effectiveness.

In the CAA report, the first in the cause list is the crew did not follow the requirements of the minimum equipment list (MEL) when taking off. Obviously the CAA thinks this is the most important one for the accident. The MEL is a list required by the manufacturer which shows under a normal flight condition what devices and equipment or systems of a plane must be able to maintain regular operation, and also what could be tolerated to malfunction or even be out of order without loss of airworthiness. Hence, the MEL is what should be strictly mandatory since it directly relates to the airworthiness of the aircraft. According to the CAA report, we knew that the cockpit voice recorder indicated the right main bus wasn't functioning, and the crewmembers understood this malfunction would cause some important systems on the airplane to break down. However, the report also mentioned that the crew was not aware that the right main bus malfunction would cause many important systems to fail. What is more intriguing is that the report at last said that the crew completely understood that the plane took off in an un-airworthy condition.

So here come the questions: in the CAA report, these contradictory statements were listed all together:

- The flight crew did not comply with the requirements in the minimum equipment list to take off.
- The crewmembers understood the malfunction of right main bus would cause some important systems on the airplane to break down.
- The crew were not aware that the right main bus malfunction would cause several important systems to not function.
- The crew completely understood that the plane took off in an un-airworthy condition.

The above statements clearly clash with each other, yet one thing is for sure: the crewmembers knew that the right main bus wasn't functioning before take-off, and they also knew that many systems would break down as well. About this, we could further reassure ourselves from the cockpit voice recorder. Second, judging from the highly professional training for a pilot, it's almost impossible that the pilots would not know how severe the situation was. Therefore, it is very logical to conclude that the crewmember completely realized that the airplane was not airworthy for flight. Furthermore, on the flight manual of the company, it clearly showed that the captain must make sure the plane is airworthy as determined by the minimum equipment list.

However, it is true that the crew did take off without the airworthiness required by the minimum equipment list. Since the pilots really knew the MEL requirements weren't met, and still took off, the CAA then chose to admit that the crewmembers were not fully aware of the severity of the situation. At first look, the CAA statement seems persuasive, yet in the first page of the accident report, it surprisingly states "they knew the MEL requirements were not met and they tried several times to resolve the problem, but in vain, so they had to take off *since they had no choice*" (emphasis added). No wonder that the crewmembers still decided to take off when they knew for sure that the plane was not airworthy: the airplane could not possibly be repaired because there was a lack of maintenance equipment at Hsinchu Airport! The crewmembers really had no choice but to fly.

Though the CAA classified the primary cause of this accident as pilot error, the pilot's error was obviously induced by something else. That is to say, the pilot's error is not the real cause of the accident, but only one event in the accident chain, although linked to the next event. In the recommendations proposed by the CAA, the very first clause is to enforce the discipline and adherence to the company's SOP. Such recommendations clearly did not point to the true cause of the accident. However, the CAA also has no choice since the classification system used by the CAA does not include any issues that would not occur from the viewpoint of Western thinking.

And where did the first cause come from? How did it happen? Why was there insufficient maintenance equipment in Hsinchu Airport? Actually, the first cause comes from outside the airport. Let us start with the history of Hsinchu Airport. Originally, Hsinchu was a military airport; the commercial air service was established on January 1, 1998. Yet, on March 18, 1998, the accident occurred, and on September 1, civil air service at Hsinchu was closed permanently. The airport lived only eight months! Establishing an airport requires considerable planning and preparation, such as airspace planning, land acquisition, buildings, transport, connecting roads, market forecasts, customers, air routes, flight scheduling, etc. It is an extremely complicated task. Yet the establishment was so unimaginably rough and so unprofessional that it took only three months to trigger an accident and eight months for the airport to shut down.

In order to set up the civil air service at Hsinchu Airport, local political figures fought for more than 10 years, yet in vain. In Hsinchu there was basically no demand for civil air service other than from politicians. However, in September 1997, the situation changed drastically and this case gained a rapid advance. The reason was that the general election for the president of Taiwan began in 1996. The candidates had to respond to local voters' requests to get elected by issuing blank checks to them. Under tremendous political pressure from the president, the CAA then spent only three months preparing for the opening of the civil air service at Hsinchu Airport and began operations on January 1, 1998. At that time in the original air force base, there was no appropriate building for terminal service, so the CAA had to use an old building, 3 kilometers away from the runway, originally occupied by the air force engineering team for the temporary space to accommodate passenger check-in. Passengers would check in at the old building and then be transported to the runway by buses rented by the airlines. According to the director of the airport, this situation would last for about five years. And when the airport first started to operate, it could only be used during daytime since there was no lighting equipment. This casual attitude to the establishment of this airport is so unbelievable.

Hsinchu Airport started operating the route from Hsinchu to Kaoshiung on March 16. And the accident took place only three days later. Actually, the airline originally was not allowed to fly the Hsinchu–Kaoshiung route since its record wasn't good. Yet the CAA granted the airline a certificate for chartered flights for the route on 16, 17, and 18 March. Unfortunately, the airline had not even completed the first three days when the plane crashed. There is a well-known saying: if something can go wrong, it will. Actually, it talks simply about basic probability. Take a very simple example: if the probability of an event happening is 1%, in 100 trials with the given conditions provided, it is probable that the event will happen at least once. The tragedy of Hsinchu happened after only three days of flying; we can roughly calculate the corresponding percentage of the occurrence of this accident combining the given conditions from the company and the airport. We assume that the airline operated 10 flights per day, so in three days' time, they would have flown 30 times. And within these 30 times, one accident happened, so the sampled probability of the plane crash is 1/30. Such an incredibly high risk created altogether by the airline, the CAA, the airport and with the public included is placed on the shoulders of crewmembers. Wouldn't this be too heavy?

Our society provides such a high-risk flying environment for our pilots, and forces them to confront the dangers and traps in it. Within this environment, the safety margin for flight has been compressed enormously, with only a very limited edge left even when the plane is still on the ground. As long as there is even some minor carelessness from our pilots, the causes of plane crashes are then attributed to human error. Those answers, though, could easily blind us to the real truth, especially talking about accidents from a different culture.

When Chinese society confronts Western aviation technology with an incorrect attitude, or even does not understand the essentials of it, and still faces it with our own thoughts, our own attitudes, or our own traditional habits, the probability of tragedies taking place would definitely rise. Summing up the above discussion, the primary cause of the Hsinchu accident is not the human error of crewmembers, although they are partly to be blamed, certainly; the fundamental cause is actually the clash between Chinese society and Western aviation technology. Unfortunately, no one can find such a primary cause in the classification system from the West.

3.3 Conflict between Two Value Systems

If we take a look back at the Sanyi accident, what happened was a lack of enough knowledge about aeronautical engineering leading to ignorance of metal fatigue. The conflict basically belongs to technical or knowledge level, or, we can say, it belongs to an artifact level. As for the Hsinchu accident, the underlying cause was the faulty attitude toward Western aviation technology within Chinese society. Coming from the top of the government, the CAA, the airport, down to the airlines and pilots, almost all of them did not truly realize the basic characteristics of aviation technology, causing the Taiwanese civil aviation companies and workforce to improperly operate the unfamiliar technological system. And the Nagoya and Da-Yuang events came from the more profound cognition about flight safety of the Taiwanese pilots. The conflict between Western aviation technology and the pilots reaches deeply down to the level of the basic value systems of Western and Chinese cultures.

The Nagoya accident was initiated by the F/O inadvertently triggering the go-around mode in the final approach phase, and then induced a chain of cascading reactions resulting in the tragedy. The crucial point is that the pilots and the flight control computer fought over control of the plane: one wanted to land the plane, the other wanted to go around. This is a situation now called man–machine conflict. And the reason for man–machine conflict is thought to be the highly automated control system of the plane.

Along with the rapid development of computer technology, the giant civil airplane is getting more and more automated. The airplane itself has gradually started to develop some kind of cognitive capability. According to the design of flight control software, the airplane can actually make decisions on its own, just like a real human being. In the international aviation community, the automatic flight control system in the airplane is usually called the electronic crewmember, electronic pilot, or virtual pilot. However, an airplane is, after all, only a machine; its decision-making ability is given by the software designer, hence, it only demonstrates the will of the designer sitting on the ground. As a result, how a software design engineer conceives flight safety will be implanted into the automatic flight control software. Consequently, the highly automated airplane is just like a puppet. Though it has head, face and body, it can also speak and move, yet it is completely controlled by the hand behind. Therefore, the highly automated airplane with the cognition of a software design engineer is no longer a machine. It is an artifact, an artifact inscribed with the cognitions of Western people about technology, aviation and safety – just like Mona Lisa's smile expressed how Da Vinci perceived the beauty of a woman.

This dilemma was described by a group of French social scientists in *Face à l'automate: le pilote, le contrôleur, et l'ingénieur (Faced with Automation: The Pilot, the Controller, and the Engineer)* (Gras et al. 1994) Based on an extensive study of flight automation in civil aviation, they found that the introduction of flight management computers and other forms of automation (including datalink and air traffic management) created a new flight deck paradigm, in which autonomy of the system actually places new demands on the flight crew, including new spatial representations and new workload discontinuities, and a more abstract mode of apprehending reality (Gras et al. 1994: 77). The conflicts between the logics of the engineers who design the aircraft, the pilots who operate it, and the ground controllers who direct them are intensified in an environment of advanced automation.

According to the definition by ICAO, safety is an "acceptable level" of risk. Risk is the severity times the probability of the occurrence of an event. Every accident will cause partial if not total loss to people's lives or belongings. Therefore, so-called safety is the acceptable level of risk of loss of lives or belongings that a society can accept. It's known to all that the acceptable level of risk in each society is different. Some societies value lives very much, and even one death is a huge event for the entire society. Some do not value lives as much: sometimes hundreds of people could die in one accident, and it will bring only a very short period of noise before the cycle starts again, which causes tragedies to happen very frequently. Why is there such a difference? It of course does not just come from nowhere; it is bred by the natural and social surroundings of such a society over a very long period. And these societal traits are called "culture." Therefore, the cognition to safety in a society is actually a phenomenon of culture.

The anthropologist Mary Douglas has described the attitudes of different societies toward risk. In individualistic societies, which are exemplified in the West, risk is associated with individual misbehavior. In hierarchic societies such as China, risk is a problem of the entire society, from the Emperor on down. Thus when a device such as the airplane which is designed for one risk paradigm (stressing individual behavior and human error) is operated by a society stressing the entire society's role, some cultural adjustments are required.

So, in the fighting of human and machine in the Nagoya accident, the one that clashes with pilots is not the airplane but the Western engineer who designed the automatic flight control system. It is largely a cultural conflict issue. When the conflict took place as the plane was facing certain specific situations, the software designer expected that the pilots should behave in certain ways. However, the reactions from the pilots with different cultural backgrounds were sometimes not the same as expected by the designer, or even contradictory to it. This type of conflict can only come from the clash of the different value systems. The Nagoya and Da-Yuang tragedies are truly this type of accident – coming from a clash of the Chinese value systems with that of the West.

Whether the clash comes from the artifact level, institution level, or value system level, they basically all occur from the discord or adverse interactions between the Chinese culture and Western aviation technology. In these interactions, the reaction of Taiwan is obviously very similar to that which occurs in evolution and adaptation phenomena from biology. When Western aviation technology was introduced into Taiwan, Taiwan had to go through a process of learning, understanding and adjustment, so that she could integrate and blend this new technology into her society. This process would definitely come along with clashes at different cultural levels from shallow to deep and profound ones, i.e., coming from artifacts, institutions and the intrinsic value systems. And the bigger the cultural difference is between two societies, the more severe and deeper the clashes between these two. Furthermore, for clashes coming from shallower levels of different cultures as the artifacts and institutions, the adaptation would be easier and faster, yet as for clashes coming from deeper levels such as value systems, the adaptation would be undoubtedly very, very slow, and sometimes even impossible.

If we could accept this viewpoint, we would come to acknowledge that most of the aircraft accidents that occurred in Taiwan were due to the "improper adaptation" caused by culture clashes at different levels. In these different accidents that occurred at different stages in 40 years, these cultural clashes took place at different levels, from the artifacts to institution and then value system; the difference lies only in the depth of clash, not in the essential qualities of clash. On the other hand, the plane crashes in the West took place from "exploring into the unknown" and still remaining almost the same for all the accidents in different periods in the last hundred years. The difference was only regarding level, not the essential qualities of the accidents. Hence, although the plane crashes are the same, the fundamental nature behind them is quite different for the West and Taiwan, and it is inappropriate to discuss those accidents all together without having cultural conflict in mind.

Chapter 4
A Chinese Viewpoint of Western Culture in Aviation Safety

4.1 The Cultural Complexion of Aviation Technology

It has been more than a hundred years since the airplane was invented, yet in the entire history of humanity, a hundred years is just a blink of an eye. Despite this, the progress of the last hundred years could be considered as a great accomplishment. At the beginning of the Air Age, the first flight of the Wright Brothers in 1903 was only 36.5 meters long; though a really small step for them, it was a huge one for mankind. And, in 1969, there came the jumbo jet Boeing 747. The wing span of this giant airliner is 59.6 meters. The distance of the first flight of the Wright Brothers was shorter than the span of a B747's wing! It took only 66 years to make the first huge step of mankind – the Wright Brothers' flight – to seem really small, as the distance of their flight couldn't even traverse the wingspan of a Boeing 747!

Apart from the rapid development of this miraculous technology, aviation has transformed human society from the inside out. In just a hundred years, aviation has become the largest, widest, and most complex technological system. It has transformed the earth into a big community, and in every single place on earth, people are mutually connected. The earth has really become a global village.

However, aviation technology bears a very obvious cultural complexion. The airplane was invented in the West and, after a hundred years, aviation technology developed by the Westerners still dominates the global skies. In the global environment of aviation, every single field related to aviation, including theory, design, manufacture, maintenance, operation, maneuver, planning, regulations, traffic control, was created, designed, formulated and dominated by the West. Even the official language in this field is Standard Aviation English. Therefore, the beliefs, ideologies, thoughts and attitudes of Westerners have already penetrated into every corner of aviation technology. Furthermore, within this hundred years, in the evolution of aviation, we can barely see contributions from any other cultures. In other words, every single component of aviation technology was bred and nurtured by Western civilization. From this viewpoint, aviation is a typical Western technology. Its traits are the traits of Western civilization.

The aviation technology invented by Western civilization has thoroughly changed the world and human society. The profound influence of aviation technology has the following features:

1. **Science**: The design of an airplane has to conform with scientific laws for it to be able to fly. Scientific laws would not be changed according to anybody's subjective understanding, expectation or mood. In the field of science, there is no God, no emperor, no authority. There are only facts. Everything needs to abide by facts. Therefore, the operation of every single component of aviation technology has to comply with basic scientific requirements.
2. **Precision**: A plane needs to counteract the gravity on earth in order to fly. The unseeable and untouchable, extremely soft air, is what a plane relies on to fly. Therefore, a plane has to strictly follow all scientific laws relevant to gravity and air so that it can fly safely. That is why there are so many very precise rules in every aspect related to aviation, whether in the hardware, software, regulations, procedures, in order to make sure the system operates safely.
3. **Procedures**: In every single discipline of aviation technology, the requirement for professionalism is quite high. And the complexity level of each corresponding profession is high enough so that most of the other technologies could not compare with it. Facing such complex technology, the standard operation procedure is the best answer which can be found up until now to efficiently operate such a huge and complex technological system. Standard operating procedures (SOPs) are procedures which ensure that the required tasks are completed in a safe, effective and orderly manner based on existing knowledge and experience. In aviation technology, from flight operation, traffic control, maintenance to almost everywhere, SOPs are the backbone of all the operations. Every move of the corresponding personnel is required to follow the SOPs strictly. So, the safe and efficient operation of such a highly complex technological system can thus be achieved.
4. **Cooperation**: The aviation system is a huge and complex system, and it requires cooperation among thousands of people to be able to function smoothly. Even in the narrow cockpit, it still requires two pilots to cooperate together to safely fly the plane. Along with the airplane becoming more and more complicated, and air transport getting more and more busy and crowded, the coordination and cooperation between people from different professions has already become the fundamental belief in aviation.
5. **Documentation**: Aviation technology is an enormous and complex technological system. Once any accident takes place, the causes behind must definitely be found out in order to prevent it from happening again. Besides, an accident also represents a lesson, albeit with a high price, but with valuable knowledge and experience to be accumulated. Therefore, in aviation technology, documentation is extremely important and highly emphasized in order to keep records for possible future use. It's not only for the purpose of investigation of accidents, but also for research, development, design, manufacture and maintenance. The goal of documentation is of course the maintenance and transmission of knowledge, experience and expertise – so that aviation technology can progress day after day – as well as the prevention of similar accidents.

From a Chinese perspective, these traits are typically Western, quite foreign to Chinese. First of all, "scientific" in the Western sense isn't the appropriate word one would use to describe China, or science would have already been developed by the Chinese hundreds of years ago. Second, the Chinese language is never precise or specific: this is why the approximate Five Elements and the Eight Trigrams prevailed over China for thousands of years. Carelessness is a very obvious trait of Chinese culture. An old Chinese saying about not differentiating between the horse and the tiger is about this mentality. Besides, the Chinese are usually smart, especially smart in taking shortcuts to achieve one's goal regardless of the regulations. Those who abide by rules strictly without the cleverness to be flexible are even considered to be dumb and laughed at in Chinese culture. As for the highly emphasized procedures in aviation technology, something as foreign as SOPs have never appeared in Chinese history.

Coordination and cooperation is even farther away from the Chinese. Children have been taught by their parents for generations not to bother others' business as far as possible – since a very early time in Chinese history. Even in recent times, Dr. Sun Yat Sen still described the Chinese people as a heap of loose sand. Even entering into the 21st century, many parents still educate their children in the same way as their ancestors. Besides these very prevalent descriptions, "One monk tote two buckets of water to drink (since he had no choice but to do it alone), two monks lift one bucket of water to drink (they have to do it together), three monks have no water to drink (because they cannot decide who should be the one to carry water)" is a fabled joke known to every Chinese.

China has had a very long history of astronomical observation records, so it doesn't sound fair to comment on the Chinese people as not being good at keeping records, or passing on experiences to next generations. Yet, valuable knowledge and experience was lost one dynasty after another. Cases of having to start all over again also appear repeatedly in Chinese history. China has the most peculiar case in the development of human technology. The south-pointing cart is the first truly automatic machine in the world. It was invented the first time by Ma Juin in the period of Three Kingdoms, 2nd century AD. Strangely, the technology to build this cart was lost 10 times, and was successfully reinvented six times in history (Jing and Liou 1994) During Cheng Ho's Expedition in the 15th century, China was the only sea power in the world. Yet when it came to the end of the Qing Dynasty, China could not even find enough good ships to defend herself from the invasion of the West. At the same time, General Hwu of the Xiang Army was shocked by seeing two alien black vessels navigating along the Yangtze River against the water current all by their own power. Frightened by this incredible scene, he fell off his horse and got injured, then died several months later in deep distress. Firearms as a Chinese invention in the Ming Dynasty were already very advanced. However, at the end of the Qing Dynasty, the Chinese imperial army surprisingly used spears and swords to fight against Western rifles and guns, as if firearms had never existed in Chinese history.

Obviously, observing from China's long history, it seems that only those things that the imperial power was interested in were clearly and carefully recorded. Unfortunately, because of the dynastic cycle unique to Chinese history, the preserving of governmental records and experiences has to start over every few hundred years. As for civil knowledge and expertise, most faced relatively the same fate: civil knowledge was not kept well and intact because of the utilitarianism in Chinese culture. Even the knowledge that had been kept quite long enough, such as gunpowder and the compass, still could not be accumulated and promoted simply because ordinary people just did not have the resources to organize and improve that knowledge. Hence, gunpowder can only be used for fireworks, not for weapons or in transportation, while the compass can only be used for Feng-Shui, not for great geographical discoveries.

Western aviation technology does have a cultural complexion and clearly we know that it is not from the East.

4.2 Flight Safety from the Chinese Perspective

All technologies, including aviation, are not just useful tools. Behind every tool created lies the knowledge from which the tool was created. Knowledge is the product of a mode of thinking, and the mode of thinking is intertwined with cognition. They are all shaped by the activity patterns of the electric currents in the neuro-network of the human brain and the cultural habits of the community that produced them. Different cognitive patterns of the different things will bring about different interpretations. Different interpretations convey different thoughts, and different thoughts will then create different theories. And knowledge, techniques, management and tools come from theories. Therefore, how a society views flight safety will embrace relevant flight safety theories, and form relevant management styles, which will finally affect the performance of flight safety. The flight safety theory is a framework of explanation that is systematic and logical, and can clearly state the compositions of flight safety and the causal relations of flight accidents. It can also guide the establishment of the corresponding techniques, management, and useful tools for the operation of flight.

In the history of Western industrial safety, H.W. Heinrich had a major influence. He was the first person to propose a safety theory based on considerable scientific research into massive industrial accidents (Heinrich 1950). Heinrich published *Industrial Accident Prevention* in 1931 based on a large amount of data on industrial accidents collected from 35 years' experience of working in an insurance company. He sorted out the causal relations for 75,000 industrial accident cases: claims from insurance companies require an appraisal of real causes, so that responsibility can be placed on the right party. After in-depth investigation of all kinds of industrial accidents, he proposed the so-called "domino theory" to portray the general process of the progress of accidents, and this became the classical theory for Western industrial safety.

After thorough research of numerous industrial accidents, Heinrich proposed an elucidation of those accidents: The occurrence of a preventable injury is the expectable culmination of a series of events or circumstances which invariably occur in a fixed and logical, even if in a probabilistic (rather than deterministic), order. In Heinrich's delineation, every accident can be conceptualized as a series of sequential events. In the series, every event comes after the previous event, and triggers the next one. In other words, every event is the result of the previous one, and also the cause of the next one. The cause and effect sequence formed by these consecutive events is just like a line of dominos. When one of the dominos tips, it will trigger the next domino to fall as well, and then the next one, and so on until the last domino is tipped.

First of all, Heinrich defined what an "accident" is according to his experience. He stated that an accident is an unplanned and uncontrolled event in which the action or reaction of an object, substance, person or radiation results in personal injury. The accident comes from the effect or the reaction of some personnel, substances or energy, and causes harm to people or loss of resources. Second, accidents happen because of mistakes made by humans or because of the failure of machinery. And the mistakes made by humans or the failure of machinery come from the defects of personnel. The defects of personnel come either from the inborn inheritance or the influence from the environment.

Therefore, Heinrich listed five common factors in the accident sequence and represented them by dominos:

1. ancestry and social environment,
2. fault of person,
3. unsafe act and/or mechanical or physical hazard,
4. accident,
5. injury.

Heinrich conceived these five categories of factors would align to form a sequence or chain of cause and effect, like the dominos: when one falls it will trigger (or at least increase the probability of) the next to fall until the last one falls.

Heinrich's first domino is ancestry and the social environment, which actually means the factors coming from culture; i.e., the factors come from natural environment, societal surroundings, and the historical ones that can pass on to generation after generation. The second domino is the potential personal defects, which can be categorized as psychological factors. And the third domino represents the dangerous acts, which obviously belongs to the realm of specific behavior. Some of the unfavorable factors that cause accidents to occur most of the time originate from the deepest cultural level, and then rise onto the psychological level shaping some of the personal defects, and then surface as the specific explicit behaviors. Whenever these unfavorable factors line up together, the occurrence of accidents becomes very likely and can of course result in some mishap and bring about certain consequences.

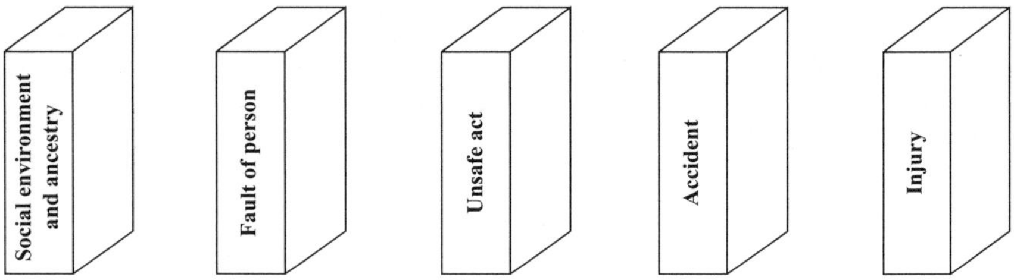

Figure 4.1 Illustration of the domino theory proposed by Heinrich

Therefore, viewing it in a hierarchical perspective, the five dominos of Heinrich represent the factors from the levels of culture, psychology, behavior, event, and consequence, from inside to outside, from obscure to obvious.

In the 1990s, Boeing proposed the accident chain theory, passing on the spirit of the domino theory (Boeing Commercial Airplane Group 1994) (see Figure 4.2). The theory conceives the happening of accidents as a chain of events aligned as the causes and effects with each other in a sequential manner. In other words, the accident chain theory visualizes the aviation accident as being a sequence of events formed by the chain reaction in which each link represents an actual incident or a mistake resulting from the previous one and also resulting in the following one in a logical way. From this perspective, the spirit of the accident chain theory and the domino theory are basically the same. However, there are still differences between them. The domino represents the conceptual category, and there are only five of them. The link in the accident chain portrays actual fact or concrete behavior, not limited to just five categories. It could be any event from the crew, traffic control, airport, weather, airplane, or maintenance, etc.

In 1990, Professor James Reason from the Psychology Department of the University of Manchester proposed the "Swiss cheese" theory, a defense in-depth concept, from an organizational perspective (Reason 1990).

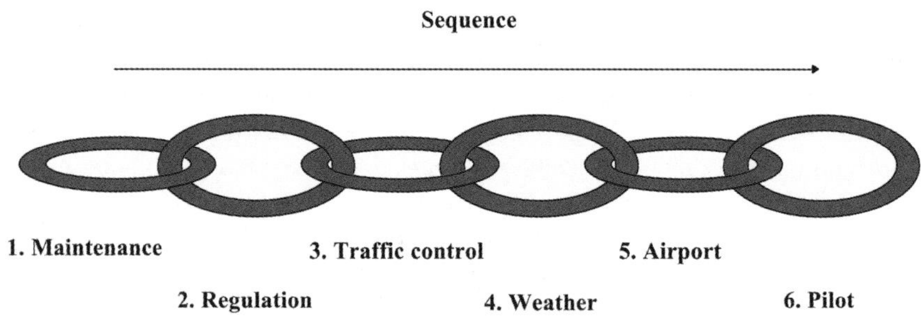

Figure 4.2 Illustration of the accident chain theory proposed by Boeing

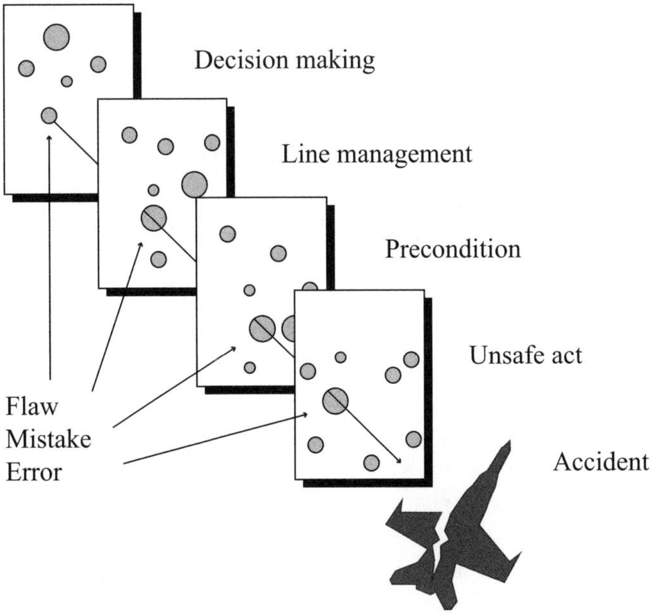

Figure 4.3 Illustration of the basic concept of Reason's "Swiss cheese" theory

Reason used several slices of Swiss cheese, known for its irregular holes in each slice, to represent the dynamic process of how an accident happens. Each slice of cheese represents a stage of an accident, and each hole on a slice of cheese represents the error or mistake of that stage. In most situations, the holes do not line up, and the error chain is broken; when an accident happens, the accident itself is like a ray of light that passes through the holes of the slices of cheese. When an error takes place, the optical light goes through that slice, and when the optical light goes through all of the slices, the accident takes place (see Figure 4.3).

Each slice of cheese represents a different stage of an accident. These stages correspond to the different phases of the industrialized mass production system. For every such production system – such as mechanical, chemical, electrical, civil, nuclear industries, including the aviation transportation system – there are five basic constituent phases as described by Reason: (1) decision-makers; (2) line management; (3) preconditions; (4) productive activities; (5) defenses. These five production phases form a sequence from top management down to production with feedback loops between each of them.

In the five production stages listed above, there could be flaws, errors and even mistakes throughout. Reason used holes in cheese slices to represent every possible flaw in every stage, for example: (1) fallible decisions; (2) line management deficiencies; (3) psychological precursors of unsafe acts; (4) unsafe acts;

(5) inadequate defenses. In most cases the holes do not line up, but when they do these different flaws in each stage form the window of opportunity for the accident to occur, and once the flaws, errors or mistakes connect together, the accident happens.

The three popular theories for aviation safety described above share a common perspective: they all conceive flight safety from a sequential point of view. That is to say, all three theories build the explanation framework by putting in place the factors involved in the accidents in a sequential, cause-and-effect order. The only difference is the depth of the explanation. In domino theory, the first domino represents the cultural aspect inherited from ancestors. In the accident chain theory, the link in the chain represents specific and observable facts. As for the "Swiss cheese" theory, the depth of its explanation is just between the above two. It does not go as deep to the cultural level, nor does it just point to specific facts; it digs only to the decision-making and management level of an organization.

In these three theories, the domino, links or cheese slices only represent a stage of causes. And, these causes are aligned in a cause-and-effect order, with every one being the cause of the next and also the effect of the previous. Therefore, we can depict these as the "sequential causal relation point of view" type of aviation safety theories.

Why do Westerners all examine aviation safety in a sequential way? Of course, there must be a reason for that. Actually, it is a very deep reason.

First of all, why Westerners think in this way is obviously influenced by the process of accident investigation. There are three main focuses in the accident investigation: they are findings, causes, and recommendations. Findings are those facts which are significant to the accident. Causes are those findings which are causal to the accident. Recommendations are the suggestions of the investigative body as to what should be done to prevent recurrence of a similar accident (Wood and Sweginnis 1995).

The first step of accident investigation is to discover all of the factors that are significant to the occurrence of the accident. These findings are not necessarily the causes of the accident. And most of the time, most findings are inter-related with each other. Usually, all the findings are arranged according to the order of occurrence or the sequence of timing to describe the accident so that people can understand it easily. The findings of an accident are normally listed chronologically, thus, it will be more helpful to identify the causes of the accident.

As for cause, in the Annex 13 definition of ICAO, it represents the actions, omissions, events, conditions, or a combination thereof, which led to the accident or incident. The Federal Aviation Administration of the United States then defines cause as something that brings about an event, or a reason for an action or condition. But, there are also many other definitions of a cause. For example, the actions, errors, conditions, environment or the combination of them to trigger or maintain an unfortunate chain of events, a cause could be a single event or a series of chain events together causing the accident to occur, or influencing factors in the causal chain causing the accident, or the findings maintaining an unfortunate sequence of events, or findings which sustained the mishap sequence. In addition to the above, there are many different kinds of causes, such as probable cause, primary cause, contributing cause, main cause, direct cause, underlying cause or root cause, etc. Therefore, the definition as well as the concept of a cause is still debated and not clearly settled or agreed by all parties today. Even whether or not it is appropriate to use the word "cause" to describe the occurrence of an accident is still in discussion.

Even if there are still many different definitions of a cause, we can still find that what is controversial is the definition of a cause, in other words, the cognition about causes are different for different people. As for arranging all the findings relevant or causal to the accident in a sequential order, this is not an issue at all.

In the actual operation, since the findings that are causal to the accident are to be aligned in a sequential order of occurring, after collecting all relevant evidence and findings, the investigation team must start finding the causal relation from the last finding of an accident, i.e., the accident itself in a reverse order up to the first cause of the accident.

Take the Nagoya accident as an example. The last event, the final link in the chain of this accident, is the plane crash and 264 fatalities and 7 seriously injured. After the last finding listed, in order to find all the causes, it is necessary to follow the sequence of the cause and effect step by step upward until the root cause found, so that the chain of causes can be revealed. Why did the Nagoya accident occur? Because the airplane stalled and fell to the ground. Why would the airplane stall? Because the airplane pitched up significantly. Why would the airplane pitch so much? Because the alpha-floor protection was activated. Why would this protection be activated? And so on and so forth in a similar fashion, until the first cause deviating from the normal situation was found. In Nagoya, the first cause was the inadvertent trigger of the go-around lever by the first officer. After all the relevant findings, no matter what we call them, are basically aligned in the cause and effect order, the accident chain with every link is then completed. Once all the events and their causal relations are clarified, the accident investigation work is basically completed.

If the investigation delves into any connecting point chosen from all the findings of actual facts, the investigation crew would right away find themselves lost in an ocean of facts without any clue. Therefore, in the field of accident investigation, there is a saying that goes, "The way to investigate an accident is to begin at the beginning or the end, never start in the middle" (McCormick and Papadakis 1996).

According to this basic thinking, people would naturally think that the reason for the accident comes from the specific connection of a chain of flaws, errors, or mistakes under the right conditions. So, if an explanation framework is to be built, it will be very natural to align all those facts in the order of causes and effects sequentially, and further illustrate them. That is how those sequential causal relation point-of-view theories for flight safety such as the domino theory, accident chain theory and Swiss cheese theory came to be formed.

Flight safety theories were originally established to provide a framework of explanation to guide accident prevention. It is natural that the basic thinking of it is influenced by accident investigation. However, this set of basic thinking that Western conceptualization relies on for accident investigation was not just formed after the time of industrial accidents in history. Western civilizations have long been reasoning from the point of view of sequential causal relationships, even including historical events.

The Japanese historian M. Watanabe once did a thorough investigation of how differently pupils (university students) and their teachers (professors from Japan and from the United States) view historical events. He found that Japanese people tend to talk more about the situations and contexts of historical events, whereas Americans tend to talk more about the important causal relation of historical events (Watanabe 1998). In class, Japanese teachers would start from describing the important background elements, including the situations and detailed contexts of a historical event, and then chain them up in a time-sequential order. Japanese teachers would encourage the students to put themselves in the historical figure's shoes to do analogy of the situations that they were facing, so that they could understand the mindsets and the emotions of the historical figures and so that the students could empathize and explain the behaviors of these historical figures to understand the evolution of historical events.

Teachers from the United States, on the other hand, spend a lot less time on the background of any historical event. They very often start from the outcome of a historical event, not from the origin or the first contributing factor. Teachers from the United States put emphasis on the discussion of causal relation of a particular event, such as "the three major reasons for the collapse of the Roman Empire." Those students who are able to point out the reasons in line with the causal relationships of such a historical event are viewed as those who have the capability of inference in history. The question "why" is raised twice as often in an American history class as in a Japanese one. Because of the huge differences of perspective, American teachers most often have a hard time understanding what exactly a Japanese student's article is about, since usually they could not find the statement about the causal relationships. And the ability of stating the cause and effect of a particular event is the most important inference ability that Americans think a student must have.

Watanabe used a term from the artificial intelligence field, "inverse inference," to describe this kind of method to find out the whole picture of a particular historical event step by step from the outcome backward to the origin. Inverse inference is exactly the product of a sequential causal relation point of view. Americans should use the exact same perspective to explain historical events and aviation accidents that are totally irrelevant; it does not happen accidentally, whereas, it illustrates how prevalent and deep-rooted such a sequential point of view is in the Western world.

Then, why would Westerners view this world in a sequential order? The main root cause of this Western mode of thinking is thought to be from alphabetic writing (Jing 2012; Logan 1986). As described by Robert Logan in *The Alphabet Effect*, the development of a logographic (word-representing) script, as contrasted with pictographic scripts, was associated in Sumeria with the development of codified law, deductive logic, and abstract science. Although Logan's thesis has been criticized by other communication theorists (DeFrancis 1984, for example), the fact that unique styles of thought are correlated with unique modes of representation of thought would seem to be unexceptionable. Around 850 BC, when the Greeks proposed the first complete alphabet, the linear mode of thinking of Westerners began. And it is as early as when the cuneiform characters of Sumerian was widely used in the ancient Middle East that people found out that the same set of cuneiforms could represent the same syllables in different languages. And, people could use only a limited number of basic cuneiforms to sequentially compose numerous ancient languages – at least 14 of the ancient languages that are recorded. After the Greek alphabet was proposed, this mode of thinking was formed and further consolidated – first in Greece, since Greek alphabet can be composed together to form more languages in a more precise way with fewer basic units.

Whether the Sumerian cuneiforms that express syllables or the Greek alphabet that expresses the vowels and consonants, they are all basic constructing elements universal to almost all human languages. Once these elements were found, they could be used to construct many different languages. The only distinction is the sequence. Western civilizations have been using alphabetical languages for more than two thousand years. If going back to the time of the Sumerian cuneiform, this mode of thinking has been in place for even longer. Because of the adaptability of culture and cognition, it is not surprising that Western people's thinking pattern has gradually been molded into a linear pattern, since the arrangements of the syllables or the alphabets into different languages all follow the same kind of lining action – in other words, like drawing lines. Every kind of language here is surely a kind of one-dimensional time series, regardless of whether it is Chinese or English. However, in Chinese, there is not any kind of basic constructing elements for written characters. In addition, there is no action such as decomposition or arrangement in Chinese as in alphabetical languages, in which the sequence is essential to those actions. Consequently, Westerners perceive almost everything with a line drawing type of thinking (Jing 2012). When this type of linear mode of thinking is applied to describing a causal relation, whether it is about flight safety or history, the cause and effect would very naturally become sequential and be put into a linear order, i.e., the causal chain. Thereafter, the emergence of the Western domino theory, accident chain theory and the "Swiss cheese" theory is totally understandable.

4.3 Aviation Safety and Culture

Every time there is a plane crash, we very often hear people talk about causes due to the airplane, weather, human error, etc. This type of categorization has been proposed by Boeing and others from experience of flight accidents over the years. There are only a few categories for the primary causes, which include the airplane, weather, airport, air traffic control, maintenance, flight crew and others.

Although this system was abandoned by Boeing in 2006, it can still provide certain useful information. As everybody knows or customarily assumes, on average 60% of the primary causes are from errors of the flight crew. Actually, if the design, management, control and operation of the whole system are all included, most accidents can be thought of as induced by generalized human errors.

In modern society, one of the first to do research about human error scientifically was Sigmund Freud. One of his monumental contributions is the concept of unconscious mind, which opened a new field of psychological research. With his psychoanalytic approach, Freud proposed that humans' mental activity operates partially under one's conscious, but most other times, it operates unconsciously. Hence, Freud believed that error was a product of the unconscious drives of the person (Strauch 2002). In his explanation of speech error, Freud proposed that virtually all speech errors are caused by the intrusion of a repressed idea from the unconscious into one's conscious speech output. Hence, people may very often say something or do something unconsciously that even they do not understand: this is the well-known "Freudian slip."

The theory of unconscious by Freud obviously has some flaws, especially where he believes that the unconscious comes from sexual trauma in one's childhood, which is really a rather narrow idea. However, no matter where we think the unconscious comes from, human error has something to do with people's unconscious: almost by definition, people to not consciously make errors. The sources of human error are most often very obscure, and the cause of human error is hard to observe and measure, even for the individuals themselves, which also causes difficulties in research.

Beyond Freud, further significant contributions to the field of unconscious were made by Carl Jung, a student and follower of Freud, although he later separated from Freud. Jung thinks that one part of the unconscious comes innately, and it has nothing to do with one's own experience, just like we all fear snakes, the dark, etc. And this part is what couldn't be explained well by Freud. Jung proposed that this part comes with the past of one nation, and it also has something to do with the evolution of the species. Jung named this the "collective unconscious," a much wider concept than personal unconscious.

From the viewpoint of psychology, what we call the root causes of a nation's general behaviors are from the past collective unconscious. Viewing from the angle of anthropology, could the collective unconscious also include the behavioral causes from the cultural aspect? Combining the personal unconscious and the collective unconscious, could human error come partly from the personal and partly from the collective unconscious? The answer is obviously yes, according to the first domino from Heinrich theory, i.e., the ancestry and social environment.

Heinrich, the pioneer in Western industrial safety research, pointed out that most industrial accidents were triggered by unsafe acts which originated from the faulty personality traits of workers. These traits usually come from the surroundings or are inherited from ancestors. Heinrich treated all of these causes from ancestry and the social environment as the first domino in a chain to emphasize its originality.

Hence, are those causes from ancestry and the social environment just the cultural factors from a different perspective?

Obviously, as early as 70 years ago, Heinrich had realized the important role culture plays in terms of industrial safety and listed it as the first cause in the chain of an accident. If it is so for an industrial accident, could it be the same for an aviation accident as well?

Boeing had conducted an analysis. They calculated the number of accidents per million departures in different regions globally according to aviation accident statistics from 1987–96. They found the accident rate is 0.5 for North America, 5.7 for South America, 0.9 for Europe, 13 for Africa, 2.3 for the Middle East, 3.8 for Southeast Asia, 2.6 for Northeast Asia, 0.6 for Japan, 0.2 for Australia, and the world average is 1.5. As shown in Table 4.1 and Figure 4.4, the accident rates for Southeast Asia, South America, and Africa were much higher than those of North America, Europe, and Australia. Why is this? If 60% of the causes of aviation accidents can be attributed to human error, such a result suggests that flight crew from these areas have worse qualities and skills.

Table 4.1 Fatal accidents per million departure, by region

Region	**Accident rate**
North America	0.5
South America	5.7
Europe	0.9
Africa	13.0
Middle East	2.3
Southeast Asia	3.8
Northeast Asia	2.6
Japan	0.6
Australia	0.2

Boeing conducted a more detailed analysis on human errors. They further differentiated human errors into about 15 actual observable behaviors (Boeing Commercial Airplane Group 1994), as given in Table 4.2. The statistics showed that there is a high percentage attributed to "not adhering to procedures" in human errors as the cause of accidents. Among all the aviation accidents around the world, not adhering to procedures as a human error showed up in almost half of the accidents. If we further classify by geography, this type of human error accounts for 41% of the accidents in the United States and Canada, 38% in Europe, 48% in Central-South America, 52% in Asia (as shown in Figure 4.5). It means that in more than half of the aviation accidents in Asia, the pilots did not strictly adhere to the SOP. And on average the percentage is 11% higher than that in the United States and Canada, and 14% higher than in Europe.

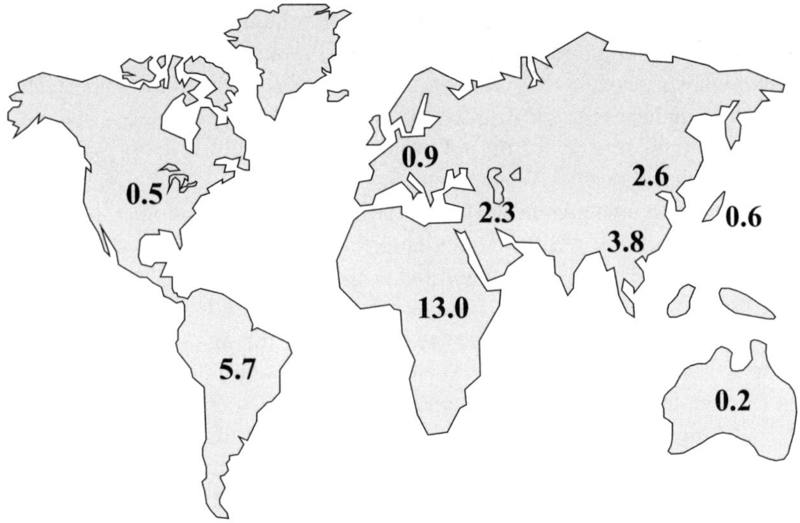

Figure 4.4 Global aviation accident rates shown as fatal accidents per million departures in different regions of the world

Table 4.2 Accident prevention strategies used by Boeing

Code	Accident prevention strategy
01	Flying pilot adherence to procedure
02	Non-flying pilot adherence to procedure
03	Flight adherence to procedure
04	Cabin crew adherence to procedure
06	Captain's cross-check – performance as non-flying pilot
07	First officer's cross-check – performance as non-flying pilot
10	Non-flying pilot communication or action
11	Flying pilot communication or action
12	Recognition and reaction to pilot incapacitation
13	Embedded piloting skills
15	Captain or instructor pilot exercise of authority
16	Use of all available approach aids
17	Go-around decision
18	Approach path stability
19	Flying pilot awareness and attention

If we judge solely from these statistics, we may come to the conclusion that Asians, comparatively speaking, tend not to like to adhere to the procedures, and that seems to be what the truth is. However, is it that simple? What is the real truth behind such a phenomenon?

Almost all the commercial aircraft in the world are designed and built by Western countries, and all the corresponding laws, regulations, procedures, and traffic control are also designed by Westerners. Even the training, operation and management are established in the West. Everything in the global aviation environment is from the hands of Westerners. When we see Asian pilots have caused more accidents because of not adhering to SOP, given the same airplanes and same environment, we could only conclude that humans are the only factor to bring such a difference. So, the difference pointed out by the statistics could only come from pilots. Therefore, every time there was an accident, people pointed fingers to the pilots' lack of training, not adhering to procedures, or even having bad habits from the air force.

However, every problem in an aviation accident is a system problem. Since it is systematic, it will be necessary to view it from a systematic angle to avoid bias. Might the previous data point the other way around? To say that it is the plane Westerners designed that is not good, instead of Asian pilots? It seems that this kind of possibility is hard to rule out. When thinking about an airplane designed in the West and crashed by Asian pilots with human error as the cause of the accident, if we jump to the conclusion that it must be someone's personal errors or mistakes, without delving deep into the true cause, we might lose the chance to really solve this problem.

If we take a more unbiased, systematic approach, we will find the crucial point to this issue might very likely come from the interaction between the machine and the human. In other words, the primary cause might not be the airplane nor the pilots, but the interaction between them, i.e., the compatibility between pilots and airplanes. Western engineers would very naturally design the aviation system based on certain assumptions rooted deeply in their society. On the other hand, when Chinese pilots operate the airplanes, their behaviors must also inevitably rest on basic assumptions from Chinese society.

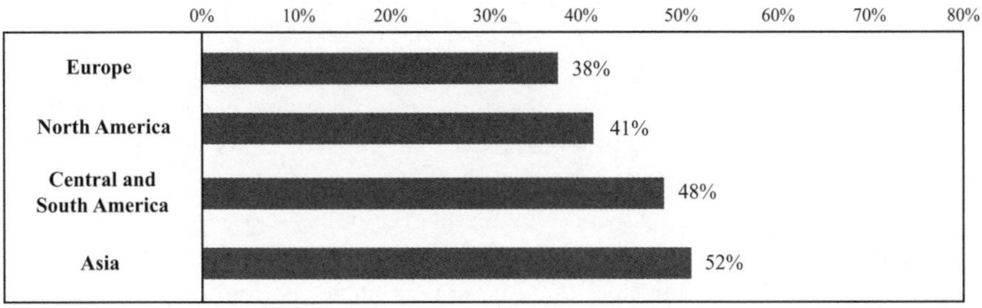

Figure 4.5 Worldwide percentages of accidents in which the human error includes pilots not strictly adhering to the SOP

It will then be very hard for these two sets of very different beliefs encountering each other in a tiny little cockpit not to clash with each other.

Most people would agree that the cause of accidents can usually be attributed to a broader sense of "human factors," including those coming from design, manufacture, maintenance, traffic control, operations, and even management. When we use the word "human error" as the cause of an accident, we have to be extremely careful. The human error implies that there is a specific person who commits some kind of error. Often a first reaction is to blame someone who people think should be responsible because they had made some mistakes. However, psychology has taught us that culture can be interpreted as the collective personality of society. When we learn that somebody made some mistakes, we should not just point to the individual person and put the blame on them. Let's first ask whether members of that society would do the same if they were in the same kind of situation. If the answer is affirmative, we have to think deeper and try to consider such an issue from the level of culture.

If one can observe somewhat similar behaviors from average members of a society, there must be an underlying reason as to why all of them behave similarly. Those who visit the United States for the first time would very likely be surprised at how much Americans seem to love to line up for everything, everywhere. Those who first visit Japan would also be quite curious about their courtesy. Why do they all act in concert without any previous arrangement? Of course, the reason is that they all accept that it is the right thing to do. That is called the value system. Everything that a society believes collectively to be good, right and beautiful, as well as bad, wrong and ugly, forms the corresponding values. The value system is the deep cause that drives average members of a society to act similarly in many ways. Since people persistently act in certain ways in certain situations day after day, even generation after generation, these original beliefs vanished from people's consciousness and became automatic behaviors. Therefore, people very often do not quite understand why they act in certain ways.

Now let's get back from the cultural aspect to review what happened in the Nagoya and Da-Yuang accidents, together with the one that occurred on New Year's Eve. In all three accidents, the weather was normal and the aircraft were functioning well, so the primary causes were all classified as human errors of the flight crew. However, when we delve deeper and try to understand the interaction between the crewmembers and the airplanes, it would be not appropriate to accredit these so-called human errors to only the flight crew's personal behaviors – because one can easily find those similar behaviors in most ordinary Chinese people. In other words, it is actually a cultural phenomenon.

From the interactions between crewmembers, the most obvious phenomenon came from the operation of power and authority gradients. In the narrow cockpit, to accomplish the flight mission, there must be a clear distribution of tasks and a hierarchy of positions since there must be one final person to make decisions at crucial moments. Therefore, there is a very clear power structure in the cockpit. One person plays the role of the superior, the senior and the teacher in the cockpit; while the other plays the role of the subordinate, the junior and the student. The activities carried out in the cockpit can be seen as a kind of movement of power operation. Different countries in the world, under the influence of different cultures, would have very different power operation patterns and different authority gradients, and very naturally would have different effects on how the crewmembers behave in the cockpit. Therefore, how could the performance of flight safety in different regions worldwide be the same?

In 2001, Jing conducted a preliminary statistical analysis on the effect of the power operation on flight safety from a cultural perspective (Jing et al. 2001). He found that the correlation between them is high enough and could not be disregarded. In the academic community, there exist several global databases about culture variation. One of them, from the perspective of comparative psychology, was created by the Trompenaars Hampden-Turner Intercultural Management Group. They collected over 50,000 surveys from more than 100 countries with some 30,000 cases valid in a cross-cultural study of business staff around the world to establish a very thorough and complete database to provide management consulting for multinational enterprises (Trompenaars and Hampden-Turner 1993, 1998). In their questionnaire, one question was about order and obedience between boss and subordinate:

> A boss asks a subordinate to help him paint his house. The subordinate, who does not feel like doing it, discusses the situation with a colleague.
>
> 1 The colleague argues: "You don't have to paint if you don't feel like it. He is your boss at work. Outside he has little authority."
>
> 2 The subordinate argues: "Despite the fact that I don't feel like it, I will paint it. He is my boss and you can't ignore that outside work either."

Figure 4.6 shows the percentage of those who said they would not paint the house for their boss, broken down by different countries. Surprisingly, among the 52 countries that Trompenaars listed, the one with the lowest percentage is China (32%) – only less than one-third of the subjects would not paint the house for their boss. In other words, on average, two out of three people would paint the house for their boss. For the United States, the percentage is 82%, that is to say, on average 18 out of 100 would paint the house for their boss simply because he was the boss. The country with the highest percentage among all 52 countries is Sweden (91%) – almost no one would paint the house for their boss.

Also given in Figure 4.6 is the average accident rate of these countries from 1981–2000. The basic correlational study can thus be done with these numbers. Regardless of the different combinations of the questions being answered in the questionnaire, the correlation is always higher than 0.6 with the accident rate. As everybody knows, high correlation does not necessarily mean there's a causal relationship. However, as the question was designed to measure the attitudes of different countries toward how power is manipulated between the boss and the subordinate, comparing this with the operation of power between the captain and the first officer in the cockpit, the existence of some kind of causal relationship between them, given such a high statistical correlation, can thus not be easily ruled out.

Obviously, the percentages in Figure 4.6 reveal a cross-section of the culture of authoritarianism in different countries. In the East, especially China, people emphasize respect and obedience to the boss, while in the West, people comparatively accept that every human is created equal. In the West, authority tends to be formal/bureaucratic, in Max Weber's terms, whereas in China it is much more personal/patrimonial. It is quite surprising to find that there is a correlation as high as 0.6 between the accident rate and the attitudes that different countries take when facing authority. Furthermore, the measured attitudes come from the database of the Trompenaars Hamden-Turner Group, a management-related study that has nothing to do with aviation accidents. Consequently, the meaning of the 0.6 correlation is far-reaching; at least, it cannot be fully comprehended by just using a simple concept such as human error.

Country	Percentage of not painting the house for the boss	Accident rate
China	32	2.96
Singapore	58	1.11
Indonesia	58	3.25
South Korea	65	2.83
India	66	3.45
Thailand	69	1.83
Japan	71	0.39
Brazil	77	2.30
USA	82	0.30
Germany	83	0.35
France	88	1.13
UK	88	0.35
Switzerland	90	0.55
Sweden	91	0.00

Figure 4.6 Percentage, by country, of those who would not paint the house for their boss and correlation to accident rates, 1981–2000

From a cultural point of view, a common and specific collective behavior from a group of people can be thought of as a cultural phenomenon, because they are all driven by the same underlying values. Therefore, the driving value behind the power operation between the captain and the first officer has become a very important subject in research about global flight safety. The specific behaviors are what can be seen from outside and are very easy to observe, study and categorize. The underlying values cannot be observed and are not that easy to understand, never mind studying and measuring them.

Values and their measurement, of course, have been studied for many years by psychologists, sociologists, and anthropologists. There are certain fundamental difficulties in the research about values. From an operational viewpoint, values produce automatic rules and routine when repeatedly facing some specific problems. Automatic response is the basic operational trait of value. Because it is a process of automation, it is very difficult to get a clear explanation from the person being studied. This is because even the subjects themselves do not know why they would act in such a way. Therefore, any kind of value research can only be conducted by observing the corresponding apparent behaviors – either through observing in natural surroundings or observing through questionnaires – and then proposing an interpretation from the researcher's angle to approximate the target value. Even so, research into values – no matter how hard the researcher tries otherwise – is always flawed, biased, limited, and incomplete, just like a blind man who hopes to know exactly what an elephant looks like by only touching it. And it also makes the very limited value studies appear as if they are just telling their own stories. This is the main difficulty that every cross-cultural research study has to face.

Let's assume that the questionnaire narrates a situation where a cultural value is expected to be observed. The subject chooses how he would react accordingly, and then the researchers try to find the beliefs behind the reaction and interpret them. In this process, there is a difficult obstacle to overcome. Even when the researchers know that most people would behave in a certain way under certain conditions, they still have to speculate about the underlying values. And it is well known that human thought is basically the product of culture, since it is bred from ancestry and the surroundings, including one's family and education. Therefore, any scale used in cross-cultural study for measurement will surely come from the specific culture to which the researcher belongs. In other words, the so-called cross-cultural study is undoubtedly using the scale of a particular culture to measure another culture, i.e., it is always subjective. It would be fair to say that there is no objective cross-cultural study in this world because there is no objective measuring scale. Hence, all cultural research results from alien cultures can be accepted only as a reference.

In the academic and aviation communities, people frequently quote the four-dimensional value system created by Geert Hofstede as a basis for talking about cross-cultural issues (Hofstede 1980). Hofstede twice conducted a large-scale investigation within a multinational corporation to collect 117,000 questionnaires from subsidiaries in 66 countries between 1967 and 1973. The questionnaire tried to explore the composition of culture from a comparative psychological perspective.

According to these results, Hofstede proposed a four-dimensional structure to illustrate the value systems of different countries. Many years later, this four-dimensional system has become broadly accepted within human factors research.

The four dimensions Hofstede proposed are: power distance; individualism; uncertainty avoidance; and masculinity.

1. **Power distance**: In its basic meaning, power is the ability to decide others' behavior. Power distance means to what degree the power is unequal. In any organization, some power distance is required, or else the organization would not function. Within organizations, power distance normally appears in the hierarchy between boss and subordinate. When both the boss and the subordinate accept some kind of unequal relationship, it represents the corresponding value or belief regarding power in this organization. Therefore, the power distance is relevant and appropriate as a measure of a society's value system. According to this, Hofstede defined a power distance index to describe the extent of the difference between how the boss can decide the subordinate's behavior and how the subordinate can decide the boss's behavior.
2. **Individualism**: Individualism is about the relationship between an individual and a group of people. Humans are social animals, but their status can vary from society to society. Individualism has something to do with the range of self, the meaningful territory of one's existence. In Western thought people accept that the existence of an individual can be separated from a group, and the meaning of an individual is defined by oneself. How people think about the relations between themselves and others will shape the construct and operation of the interpersonal relation in a society, and would definitely mould the personalities of the individuals in that society. Therefore, the perceived relationship between individual and group in a society must be closely related with the value system of a society.
3. **Uncertainty avoidance**: Life is full of uncertainties. People tend to have anxiety, worry and even fear of uncertainties encountered in life. Consequently, people created some means to try to avoid the uncertainty or to build certainty. Usually, people deal with uncertainty in three different ways: technology, laws/regulations, and religions. Technology helps people to face uncertainty from nature, laws and regulations help deal with uncertainty from other people, and religion handles the uncertainties in life that people are not able to face. How people treat these uncertainties represents certain values in a society.
4. **Masculinity**: The natural difference between male and female is a fundamental issue that every society has to deal with. Whether the difference from biological nature also means there must be a clear discrimination of their social roles varies from society to society. The two sexes are bestowed with different values and expectations from their very early ages. In the socialization process from families to schools and society, the differentiation

is further consolidated. In many cultures, the male is expected to be brave; the female is expected to be gentle and soft. Naturally, how people perceive this issue shapes the corresponding values.

The distribution of Hofstede's four dimensions among selected countries (USA, Sweden, China (Taiwan), and Japan), is shown in Table 4.3.

Table 4.3 Hofstede's cultural dimensions, selected countries

Dimension	USA	Sweden	China (Taiwan)	Japan
Power distance	40	31	58	54
Individualism	91	71	17	46
Uncertainty avoidance	46	29	69	92
Masculinity	65	5	45	95

The four dimensions that Hofstede proposed are inter-related. For example, those who value self-accomplishment, self-dependence and self-reliance tend to rely less on their boss or other authority. Therefore, a society with high individualism would very naturally have a lower power distance.

In the aviation community, there also exists a cross-cultural database. A team from Professor Robert Helmreich of the University of Texas designed a cross-cultural questionnaire for commercial pilots around the world, the Flight Management Attitudes Questionnaire (FMAQ) (Helmreich and Merritt 1998). The purpose of this questionnaire is to reveal the attitudes of pilots about order, communication, workload, automation, organizational climate, and values of work. From 1993 to 1997, the released and returned questionnaires included over 15,454 male pilot respondents from 36 airlines in 23 countries. The Helmreich group analyzed the data and then presented in detail the influences of professional culture, organizational culture, and national culture on the activities of the crewmembers in the cockpit. The database is the most comprehensive one about the influence of culture on aviation up to now.

According to Helmreich's research results, the strongest differences between pilots of different countries were observed with regard to:

1. command roles and interactions;
2. attitude toward orders, rules, and routines; and
3. preference for and reliance on automation.

For the first item, Helmreich designed several questions based on the concept of power distance proposed by Hofstede, hoping to measure the power distance between the captain and first officer in the cockpit for different countries. The results can be synthesized into a score, which indicates the attitude difference when pilots face authority. A lower score denotes a lower power distance; a higher score signifies a higher power distance. Judging from the scores, pilots from Europe and America have lower power distance; in contrast, pilots from Asia possess a higher power distance, as shown in Table 4.4.

Table 4.4 Helmreich's scales, selected countries

Scale	USA	Sweden	China (Taiwan)	Japan
Command roles	46	46	90	68
Rules and orders	44	51	78	57
Preference for automation	58	48	73	n.a

The second item is about the attitudes toward orders, regulations, and procedures. However, if we judge from the original text in Helmreich's questionnaire, what he really wanted to measure is the extent of blind obedience the pilots have to orders and regulations. And the results showed no surprise. The score is higher for most pilots from Asian countries, while the score is lower for most pilots from the West. The results coincide with general understanding, because the tendency of obedience for Westerners to any form of authority, whether it comes from an individual or order, and regulation is supposedly lower.

The third item is about the degree of reliance on and trust in automation. Here, the automation represents the programmable computerized flight management system. Pilots from Asian countries comparatively prefer to use the automation to help their flight, while pilots from the West usually do not. A higher score from the questionnaire means higher preference to automation. Compared with Asians, Westerners evidently tend to trust the automation less.

Superficially, the databases of Helmreich and Hofstede are completely different. Helmreich's research objective is to understand crewmembers' attitudes toward resources management in the cockpit, and placed emphasis on whether crewmembers' behavior is effective or not for implementation of cockpit resource management, without touching the values behind the behavior. In contrast, Hofstede's research emphasis is to search for values in different cultures, and he further brought about a four-dimensional value system to compare cultural differences between countries. In spite of all of these, there is a really high correlation between these two databases.

The correlations between the three differences in national culture by Helmreich and the four dimensions by Hofstede are as follows. The first item has a 0.79 correlation with power distance, and a 0.86 correlation with individualism; the second item has a 0.67 correlation with power distance, and a 0.88 correlation with individualism; and the third item has a 0.74 correlation with power distance, and a 0.74 correlation with individualism. These three items are less correlated with uncertainty avoidance and masculinity, as shown in Table 4.5.

Table 4.5 Correlation of Hofstede's indices and Helmreich's scales

Helmreich's scales	Hofstede's indices			
	PD	IDV	UA	MAS
Command roles	0.79	0.86	0.51	0.12
Rules and orders	0.67	0.88	0.47	0.10
Preference for automation	0.74	0.74	0.63	0.22

The correlations between the three items observed by Helmreich and power distance are 0.79, 0.67, and 0.74. The correlations between the three items with individualism are 0.86, 0.88, and 0.74. Each of these correlations was higher than 0.6. If we compare the correlations higher than 0.6 between authoritarianism and global accident rates from a previous study (Jing et al. 2001), it will be quite surprising to find that no matter how the correlation was calculated, a value higher than 0.6 always appeared. It is very unlikely that this is pure coincidence, and there must be something very fundamental worth figuring out in it.

According to the research by Helmreich, among the pilots coming from different countries, the Chinese – including Taiwanese pilots – ranked high in all of the three items about the national differences of attitude in the cockpit. Moreover, for the first item, China was ranked first, for the second item it was ranked third, and for third item it was ranked fifth. Therefore, judging from a Western point of view, the power distance and the collectivism, as opposed to the individualism, of Chinese culture is substantial and forms an intensive contrast when compared with Western culture.

However, when we try to understand and explain the behaviors of the crewmembers in major Taiwan accidents with the concept of power distance, we find that the authoritarianism of Chinese culture is extremely complex; authority cannot be interpreted simply through the concept of power distance. Authority is not nearly as uni-dimensional as Hofstede's scale might suggest. Power distance delineates the difference between the abilities of how well a boss can decide the subordinate's actions and how well a subordinate can decide the boss's actions. The power distance of Chinese culture is huge according to Helmreich, which means that the boss has great power in deciding how subordinates act. In contrast, the subordinate has almost no power in deciding the boss's action. Can such a concept of "distance" express fully the interaction between the captain and first officer in the Nagoya accident? The first officer tried many times to deactivate the go-around mode after he inadvertently triggered it, yet in vain. Why didn't the first officer try to seek help or speak out? Why didn't the captain himself ask? Why did the captain do exactly the same after taking over the control from the first officer at the last moment?

Similarly, in the Da-Yuang accident, the captain consecutively made many mistakes, yet the first officer from the beginning until the end only obeyed without question. Even if the first officer did not have the power to alter or question the captain's actions, couldn't he possibly just try to remind the captain? In the New Year's Eve accident, the flight attendant sat in the cockpit for the whole flight, and twice she interrupted the communication between the flight crew and the controller. Why didn't the captain, who actually had the power to decide how his subordinate should act, correct the flight attendant and just let things like this happen? Why didn't the captain take over the control of the plane when he realized the first officer was operating the plane incorrectly? Why did the captain even call the first officer "sir"? Could we explain these types of strange behavior through the concept of power distance?

Why can't the concept of power distance be used to fully explain Chinese authoritarian culture? The reason is very simple. Chinese people traditionally never conceive the operation of power as a "distance." Western people tend to think in terms of drawing lines to interpret the world because their mindsets have been shaped by their alphabetical writing over thousands of years. For example, they would form the concept of a causal chain by drawing lines through various factual findings to figure out the hidden causal relations in aviation accidents. Westerners also comprehend the operation of power with line drawing and come up with a concept of distance in describing power. This concept of "distance" is portrayed schematically in Figure 4.7.

However, the Chinese way of thinking isn't anything like the linear type; instead, it is an ideographic type. Since the Chinese do not use the alphabetical writing system, and there is no alphabet in the Chinese language, this means that the Chinese do not differentiate the sound in details. Consequently, the Chinese language becomes a kind of ideographic one that primarily uses graphics to convey messages instead of sound. Hence, it will be quite reasonable to treat the Chinese characters as abstract hieroglyphs. The Chinese characters originated from the oracle bone inscriptions which had been in place for more than 3,500 years. Due to the plasticity of the human brain, Chinese mindsets have been molded into a completely different ideographic type of thinking compared with that of Westerners. Therefore, when thinking, Chinese people tend to draw figures instead of lines. When one is drawing pictures, the order of drawing is definitely never important, even meaningless. So, it is very hard for the Chinese to establish linear concepts such as sequence, distance, and even logic.

The ideographic thinking shapes Chinese minds to think in a way that is just like drawing a whole picture. It creates a way of thinking that emphasizes totality more and creates the systemic view involuntarily, and highlights the dynamic interactions between the components within it. The Chinese believe that as long as all components interact positively, that is to say, stay in harmonious equilibrium, the system as a whole can then grow and develop well. Once the equilibrium is lost, the system would enter into a vicious cycle and deteriorate. Therefore, the Chinese tend to worry instinctively about losing equilibrium and try to avoid destroying the positive interactions between components, thus forming a major Chinese cultural trait of emphasizing harmony wherever possible. Consequently, the Chinese most often would still perceive the interactions between the boss and subordinate in the cockpit from the totality-equilibrium angle, emphasizing harmony, with their ideographic mode of thinking.

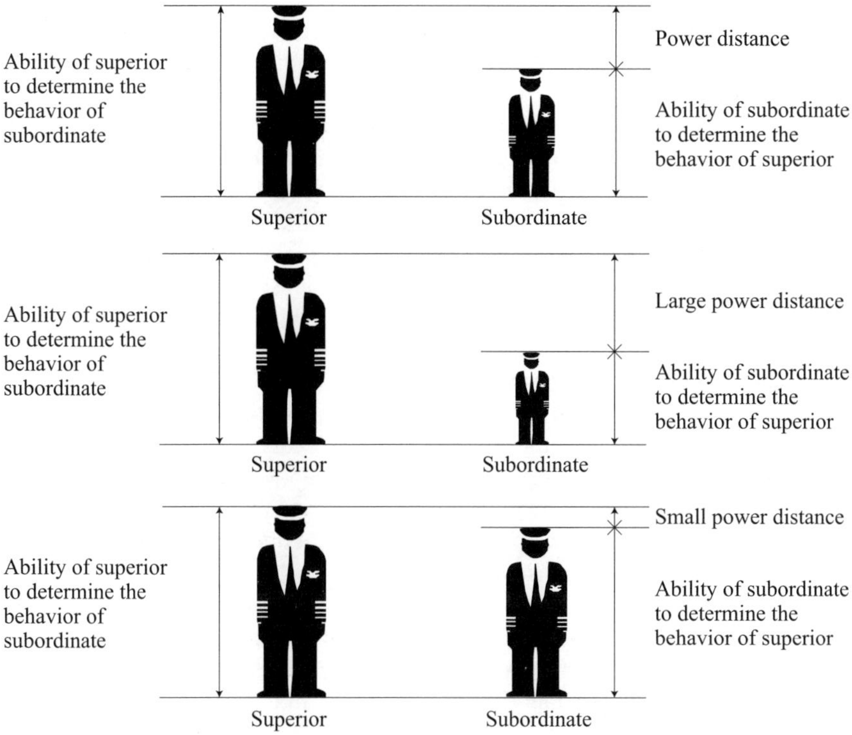

Figure 4.7 Westerners comprehend the operation of power with line drawing and come up with a concept of power distance

Being Chinese, when the subordinates face the boss, they would unconsciously show an obedient attitude and possess a feeling of fright, out of fear of the power the boss holds, since the subordinates normally would not know the real intention of the boss. However, the subordinates, on the other hand, would expect the boss to protect them instead of punishing them, or even further to bring benefits to them from the power owned by the boss. When the boss deals with subordinates, by the same ideographic thinking, the totality-equilibrium perspective, the sense of responsibility would automatically be aroused. When the boss enjoys obedience and trust from the subordinates, the boss will naturally feel that they have the responsibility to take care of the subordinates and to meet their expectations. These kinds of interactions and balance between boss and subordinate cannot be fully described with a concept as simple as "power distance."

Individualism involves one's perception of the extent or reach of self, that is, where the boundary of meaningful existence of self lies. In the West, people are used to dividing the interpersonal relationship into two categories: individualism and collectivism. It is a typical dichotomy from Western linear thinking. Westerners classify whether a culture belongs to individualism or collectivism based on where the boundary lies in terms of individual and group interaction. In other words, they observe interpersonal relationship from an angle of how big the distance is from the boundary to the end of the individual or to the other end of the group, as illustrated in Figure 4.8. Again, in the understanding of interpersonal relationship, individualism is still a very typical Westernized concept as delineated by distance using the linear mode of thinking.

In the Tao-Yuan accident that occurred on New Year's Eve, why would the crew allow a flight attendant to sit in the cockpit? What kind of driving force made them violate the rule? Why didn't the flight crew stop the flight attendant yet reply to her questions first when she caused definite interruptions to the communication between the crew and the controller twice? And, why didn't the captain, acting as the boss, take over the control of the plane when he realized the mistake of the first officer, who was supposed to be the subordinate? What kind of interpersonal relationships would bring about this kind of strange interaction? Can we comprehend this kind of interaction through the concept of individualism or collectivism? Obviously not.

To Westerners, China is a society of collectivism because the interpersonal relationships do not show the main features of those from an individualistic society. That is to say, Chinese society is definitely not individualistic. This is not an original observation, although its implications for flight safety remain to be explored.

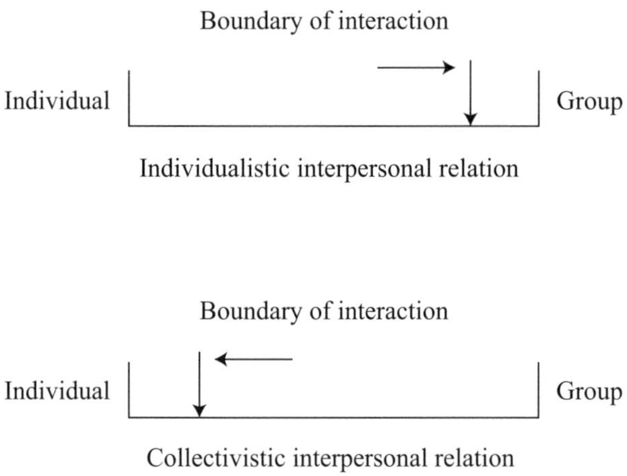

Figure 4.8 Linear mode of thinking and understanding the interpersonal relationship creating the corresponding individualism and collectivism by using a concept of distance between the individual and the group

However, in reality, Chinese society is also definitely not collectivistic. There have been many studies by Chinese academics to illustrate this (C.-F. Yang 2001b). The reason is simple. It just wouldn't be sufficient to describe the interpersonal relationship in Chinese society using the concept of individualism or collectivism, since in Chinese society the interpersonal relationship cannot be understood by any concept of "distance." First, it is widely accepted by the academic community that Chinese culture is an ethical culture. Second, the Chinese tend to use different strategies in different situations when dealing with interpersonal relation issues, a trait that is sometimes called "situationalism" (Hsu 1981). As Chinese, we will use the most appropriate strategy according to the degree of intimacy and the corresponding situation at the instance when dealing with interpersonal relations.

Hence, in Chinese culture, nothing can be completely understood with a "distance" concept. Therefore, we must be extremely careful when we use concepts such as power distance and individualism developed by Western linear thinking to comprehend Chinese culture and interpret the corresponding behaviors.

Chapter 5
Descendants of the Dragon

If we cannot understand the Chinese power operation and interpersonal relations with the concept of distance, how can we understand the Chinese, especially Chinese authoritarianism and interpersonal relations? Although the written Chinese characters do help shape Chinese culture, their influence is only on the form of Chinese thinking rather than the content of it. Similar to the West, the influence of the alphabetical language is on the form of the Western mode of thinking. That is why we call the logic created by Aristotle "formal logic." That means if we want to know the content of something such as Chinese authoritarianism, we will have to look elsewhere.

Chinese civilization is the only one in the whole world which has existed continuously from ancient times until today. Hence, there must be so many things in this culture that have an incredibly long history of evolution, starting thousands of years ago. To understand anything like this, we will have to look back in time and dig deeply into something old enough to reveal the real meaning of it. Chinese authoritarianism is such a typical example.

5.1 Traditional Chinese Authoritarianism

Culture has been an issue in the global community of flight safety for many years. There are already many discussions on how culture influences the behavior of pilots in the cockpit. Many different points of view have been raised. In the aviation literature, the most cited works about culture include Hofstede (1980), Helmreich and Merritt (1998), and Trompenaars and Hampden-Turner (1993), among others. Originally, the pilots' training and skills as well as performance were thought to be universal and standardized as they were designed. However, interaction patterns among crewmembers have a certain influence on pilots' performance. More and more studies consistently aligned the pilot performance in the cockpit into recognizable cultural traits. Therefore, the effect of national culture on pilots' performance was perceived to be more and more important. In a study directly related to pilots, Helmreich and Merritt (1998) pointed out that the strongest differences between pilots from different countries were:

1. command roles and interactions;
2. attitude toward orders, rules, and routines; and
3. preference for and reliance on automation.

Basically, all of these can originate from the mentality of authoritarianism, i.e., how people in a society face the authority which is more capable and possesses certain influences. In the present environment of highly automated flight systems, these questions of balancing the authority of the flight crew with the autonomy of the flight management systems are pressing issues all over the world (Harris 2011). When one adds cultural differences in the meaning of authority, the issue becomes quite complex. Using the database of Trompenaars (1993), Jing et al. (2001) correlated the commercial aircraft accident rates globally with the corresponding authoritarianism and concluded that authoritarianism as an element of culture cannot be neglected in global flight safety.

The meaning of authority, of course, is a subject whose discussion stretches back to Plato and Confucius. In discussions of flight operations, Jing et al. (2001) provides a concise description: "the mentality and behavior causing the distortion of individual freedom of thought and action because of superiority derived from status, resulting in a change of designed aviation operations and distribution of resources" (Jing et al. 2001). If this is an appropriate definition of authoritarianism in the cockpit, it is not surprising to find that Chinese authoritarianism is quite different from its counterpart in the West. To understand Chinese authoritarianism, we have to reach far back into history, first to the earliest contacts with European powers, and then further back to the beginnings of Chinese civilization.

In 1793, the English envoy Lord Macartney came to Tianjin, China, with the mission from King George III to visit Qing Emperor Qianlong. The aims of the mission were to put Sino-British relations upon a treaty basis, redress various grievances in the existing commercial arrangements at Canton and further ease restrictions on trade between Great Britain and China, open new ports to trade in north and central China, and, if possible, establish a permanent legation in China. The British were well aware that the Chinese considered foreign embassies as tribute-bearing missions and generally, if not always, demanded that the ambassador perform the kowtow before the emperor according to the convention. The kowtow is the ultimate ritual expressing the highest respect and obedience, with three kneelings and nine head-knockings to the ground. For the Chinese people, it is the sincerest etiquette, sometimes used at weddings to appreciate parental grace of giving birth to a child and bringing it up. Sometimes it is performed at the funerals of parents. Since the Middle Kingdom considered itself the geographic and cultural center of the world, China considered itself self-sufficient. Rather than relying on other civilizations to obtain goods through trade, China was interested in being recognized as superior to all other civilizations. Thus, trade in China was conducted much differently than it was in Europe. Since merchants had to conform to Chinese conventions, the recognition of China's superiority was a natural consequence of the tributary system.

The tributary system worked fairly well until the Europeans started showing up on the coast of China. While the Vietnamese, the Koreans, and the Japanese were more than willing to give their obeisance to the Chinese, the Europeans were not. They had no experience with Chinese culture. While they were initially willing to conform to Chinese market conventions to get their hands on valuable Chinese goods, they eventually sought to impose European-style trade. The Chinese resisted, but after many years had to succumb to Western pressure. The technological advantage that the West enjoyed allowed it to open China for trade despite Chinese reluctance to Westernize.

After respecting the Chinese market conventions for many years, the English monarchy sought to arrive at a European-style trading arrangement in China. To do this, Britain sent Lord Macartney to meet with the Chinese emperor to petition for his permission to change the market conventions governing Chinese/British trade. While they had many requests, most of these requests dealt with increasing their access to Chinese trade. They were not content with trading alongside other Europeans in the single port of Canton. They wanted access to additional ports, representation in the Imperial court, and land to store their goods.

Macartney came to the Chinese bearing lavish gifts from the British monarchy and attempting to abide by Chinese traditions. Nevertheless, his requests were completely without precedent and were all denied. As it was recorded and taught in the Chinese history, the main reason was for diplomatic protocol. Macartney was asked to prostrate himself before the throne as the Dutch and the other envoys had done before him. The ambassador agreed not only to perform the prostration but also to conform to every exterior ceremony practiced by His Imperial Majesty's subjects, if a subject of His Imperial Majesty, of equal rank to his own, should perform, before the picture he had with him of His Majesty King George III, dressed in his robes of state, the same ceremonies that the ambassador should be directed to perform before the Chinese throne. The Chinese opened the audience and, ignoring the proposition made previously by Macartney, pressed him to perform the kowtow, saying that it was a simple, meaningless ceremony, which is true for the Chinese since they had already been used to it for many years. They were willing to have some slight alterations made in the ceremony so that it was not exactly the same as that performed by the envoys of the vassal states; but Macartney would only agree to bend upon one knee before the emperor, as he did before his king. According to the British official narratives of the mission this was accepted by the emperor, and the audience took place shortly after in a tent in the Summer Palace gardens, where the ambassador was kept from before dawn awaiting the emperor's arrival. There were suspicions about whether Macartney really did kowtow or not. The ceremonial event that followed was kept a profound secret by those who witnessed it. Macartney was then given the emperor's very haughty and highly unsatisfactory answer to King George's letter. The mission failed badly in its attempt to win the trust of the Chinese authorities (Rockhill 1897).

In 1816, William Amherst was appointed as the ambassador and visited China again. Once again, he refused to kowtow and the mission failed, just like the previous one. From the conduct of Amherst, we might conclude that Macartney very likely did not kowtow as it was customarily known to the general Chinese people. However, not all envoys of Western countries behaved in the same way. In 1795 the Dutch envoys kneeled and knocked their heads on the ground and they were led into the Forbidden City, across the marble bridge to meet the emperor together with other envoys.

Obviously, the kowtow was at the center of this historical event. To the Chinese, the act of self-abasement in the form of kneeling three times and knocking the head on the ground nine times was learned in the family to express sincerest appreciation to the parents for giving life. Its significance went far deeper than merely showing respect in court ritual to the emperor. In addition, the kowtow bears implications more than an acknowledgement of the rights and obligations owed to a higher figure by a lower power as it is understood in the West. But for Macartney, the act of dropping down on both knees was reserved for begging for mercy, praying to God and proposing marriage. To kowtow to the emperor of China was an act of abject humiliation which he, as the emissary of an equally proud nation, was simply not prepared to do.

Consequently, to the British, it was quite obvious that the kowtow is something that implies total yielding to the Chinese authority. Whether they wanted to kowtow or not is surely related to the trading business. However, the final decision to yield or not should certainly come from the consideration of power. Regarding power, having linear thinking, Westerners will automatically perceive from a concept of distance to make a comparison. At that time, the British, with far more advanced technology, had already colonized much of the world. Considering power between the British and the Chinese, it would be very natural for them not to yield because of the greater power they had. "I have greater power, why should I follow your will?" This was also the supposed reason for the Dutch to kowtow since the Dutch were expelled from Taiwan in 1661 by Koxinga of the Ming Dynasty, whose descendant was later defeated by the Qing Dynasty.

Obviously, Westerners see the kowtow from an angle of power operation, while the Chinese see it quite differently. It can be agreed by all parties that to kowtow is truly a symbol of authority, the problem lies in the meaning of it. In the West, authoritarianism is related to submission to authority because of the basic assumption about power. In China, it is totally different, as can be seen from the attitude of the Qing emperor. From the very beginning, the Qing emperor declared the imperial edict to praise the British for their coming over and showing allegiance. Why did the Chinese emperor do this? The Middle Kingdom had been the geographic and cultural center of the world known to the Chinese for thousands of years; also, China considered itself to be much more advanced culturally than the surrounding nations. China had been prosperous for such a long time in history. Further, China was an empire founded on agriculture. That is to say, China was self-sufficient, not relying on other civilizations to obtain goods through trade. The Middle Kingdom needed no trade with anyone, not even the British. Even so, China was still willing to help everyone coming over. However, you have to show your submission first. If you showed your sincere attitude, of which the kowtow is the ultimate form, anything can be discussed. If you came over and did not want to show your submission, why did you come here?

In this Chinese style of power operation, useful benefits, including government income through trade, seem not to be in the mind of the emperor. The Qing emperor also appears to not be interested in doing business with the outside world. Even the comparison of power of each other was not important since the Qing emperor did not seriously investigate the military and industrial strength of the British. Compared with the West, the Chinese framing of authoritarianism is truly different.

In order to understand Chinese authoritarianism, it is necessary to know the real meaning of emperor in China. Since the first emperor of the Qin Dynasty united China in 221 BC, this system lasted more than 2,000 years, until it ended in 1911. The typical symbol of Chinese authoritarianism is the emperor. However, part of the reason for the emperor system lasting so long was because the notion of emperor overlaps largely with the image of father. Of course, this congruence of father-emperor was created on purpose, because of the need for ruling. Hence, for a long time, the images of emperor and father were entangled with each other. It becomes the so-called emperor-father in Chinese history, similar to God the Father in the West. Emperor-father is consistent with country-family. In Chinese, country is exactly country-family altogether because country is understood as a big family and the emperor is the father of the big family. Thereafter, although the typical symbol of Chinese authoritarianism is the emperor, the prototype is the father. Knowing the role of father in the traditional Chinese family is crucial for understanding the meaning of the emperor and Chinese authoritarianism in general.

According to the anthropologist F.L.K. Hsu, the dominant kinship relation in the Chinese family is the father–son relationship, while in the West it is the husband–wife relationship (F.L.K. Hsu 2000). In other words, the operation of Chinese family is dominated by father and son. In the modern West, husband and wife dominate the family. The husband-and-wife family possesses features such as discontinuity, exclusiveness, sexuality, and volition. The father-and-son family holds traits such as continuity, inclusiveness, asexuality, and authoritarianism. The difference between these two types of family is not only very large but also very meaningful. For example, discontinuity represents the fact that the relation of husband and wife will never be passed down generation after generation, while the relation of father and son will definitely go through generations because normally every father is a son and every son will be a father. Hence, the father–son relationship is continuous, like a long chain of inheritance: every father or son represents a link in the chain being somebody's son and somebody's father simultaneously.

The most distinctive difference between the father–son and husband–wife relationships is the existence of choice, or "volition" in Hsu's term. No one can choose their father, no one can choose their son, but everybody can (at least nominally) choose their spouse. Further, nobody can abandon their father or son, but it is totally different for husband and wife. In the relation of father and son, there is no right to choose for anyone. Consequently, the operation of husband–wife is basically liberal and father–son is not. In addition to this, the relationship of father–son is essentially authoritarian. Every father is older than his son. For a significant time, the father is more capable and experienced than the son. In this period, the father plays the role of giving orders and instructions, as well as being a supplier of living resources. And, the son is compelled to receive the orders and resources, like a forced follower, since the son has no choice but to rely on his father for living.

In a society with father–son as the dominant kinship relation, it can be expected that the image of authoritarian father will be planted naturally into every member without permitting anyone to make a choice. There is no need to prove the legitimacy of the father's authority. In the relation of husband and wife, it takes reason for the husband to have obedience from his wife. Because of this, the authoritarian husband is usually loathed and hated while the authoritarian father is more acceptable and tolerable. Since the dominant kinship relation will shape the corresponding value system of the society, the society dominated by the father–son relationship will normally be more authoritarian.

The typical examples that Hsu used for his study are American and Chinese families. This kind of long chain of father-and-son relationship has existed in the Chinese family for thousands of years and evolved into an indestructible pattern, often called patriarchy. In a general Chinese family, the father is supposed to be the authority and is responsible for everything. The greatest power in a family is in the hands of a father. All the other members in the family are just chosen dominated followers.

After the Qin united China, the centralized monarchy system was established and the "under-the-heaven family" political tradition was further consolidated. Since then, patriarchy served as the social stronghold to guard the ruling of the emperor. The ideology of Confucian teaching was further used as the official orthodox to strengthen the influence of patriarchy. Emperor and father were fused together to form the emperor-father, and the nation and family were also fused together as the country as well. In the long history of China, no matter how long the dynastic cycle is, this "super stable structure" (Jing and Liou 1994) was maintained until today. From then on, in every kind of organization, the leader possesses the hidden image of a father. As long as there exists superior and subordinate, the father–son relationship is always the prototype of the power operation.

5.2 Origin of the Dragon and Dragonality

Among the many characteristics of Chinese culture, authoritarianism might be one of the most important (Chu 1993). Within Western sociology, the German sociologist Max Weber is an accepted authority on the different forms of authority. Within Chinese civilization, authority as symbolized by the dragon is what the anthropologist Sherry Ortner called a "key symbol," a central narrative for explaining the society (Ortner 1973). There are already several studies related to Chinese authoritarianism. From the teaching of Confucianism, Wei (1972) concluded that the most representative character of Chinese people was an authoritarian temperament. In addition, from the viewpoint of traditional Chinese values, Wen (1972) also found that authoritarianism could explain the personality of most Chinese people. Tzeng (1972) also came to a similar conclusion starting from the personality development concept. However, most of these studies were influenced by Theodor Adorno's *The Authoritarian Personality* (1950). There were other studies revealing different points of view. From the study of many refugees from mainland China, Pye (1968, 1981) found that although those people did respect authority figures, the phrase "blind authority" could not be used to describe their behavior. On the contrary, those people did have clear expectations and judgment of the authority figure. If the authority figures were not qualified to behave authoritatively, the Chinese subordinate would resist with passive aggression. Yang (2001a) also pointed out that the Chinese do not possess an authoritarian "personality," although they do respect the values represented by hierarchical systems and elite rule. Still, it is not surprising that a group of people under autocratic rule for more than 2,000 years reveal a clear tendency toward authoritarianism, although obviously that authoritarianism might have different meanings for the Chinese. Therefore, it would be very helpful if the distinct trait of Chinese authoritarianism can be illustrated.

In a previous study, Jing et al. (2001) raised some evidence of Chinese authoritarianism in a representative case. This evidence included the fear of being punished, hiding the weakness of the authority figure, and preserving the face of authority. It was also shown that the procedural activities in the cockpit were highly vulnerable to this authoritarian mentality. However, where does this mentality come from? What is the real truth about Chinese authoritarianism? Having a better understanding of Chinese authoritarianism for non-Eastern people in commercial aviation would definitely be good for global flight safety.

Since ancient times, the Chinese emperors used the dragon to represent themselves. Obviously, this is not pure coincidence. It is surely not reasonable to say that the dragon was created to serve the needs of imperial rule. It also would not make sense to say that the development of imperial power follows the image of the dragon. However, it is very convincing to believe that in the development of the dragon – because this symbol bears a meaning that is beneficial to imperial rule and can be used to visually transmit the content of imperial power – the dragon was absorbed by the ruling class and further enriched political significance, finally forming the unique symbolic figure of traditional Chinese authoritarianism.

Moreover, after this artifact matured, its meaning was intensified dynasty after dynasty until the dragon became embedded in the minds of ordinary Chinese people. Although imperial rule was eliminated by the Republic of China in 1911, the father still dominates the operation of every Chinese family. Therefore, the implication of dragon is still inherited by generation after generation.

The dragon is the unmatched symbol of imperial power used by almost all the emperors in the long history of China (Wen 1948). Actually, "dragon" is not the correct word for the traditional Chinese dragon. To the West, the dragon, a mythological figure, is almost a synonym for evil. To the Chinese, the dragon has nothing to do with any evil. Instead, the dragon bears an implication almost exactly opposite to evil. The dragon is not a creature, not even a mythical creature. It is a very vivid symbol of Chinese authoritarianism. Reinforced by long-lasting autocratic ruling and Confucianism, the image of the dragon has permeated deeply into every Chinese mind. Today, almost every Chinese thinks of himself or herself as a descendant of the dragon. Every parent expects his son to become a dragon when he grows up. Here, becoming a dragon means being successful, i.e., wealthy and powerful. To take a popular example, the Chinese name of the famous movie star, Jackie Chan, is exactly "becoming a dragon."

There have been many stories told about how the dragon was created in China. Obviously, there is no such creature in the world, not even in fossils. Further, it is of course not a winged monster as depicted in Webster's dictionary or medieval tapestries of St. George slaying the dragon. It never existed. However, it did exist for more than 3,000 years in the minds of Chinese people. The Chinese word "dragon" first appeared in the Oracle bones and bronze in the Shang Dynasty (1751–1111 BC). If the usage of written characters is seen as the beginning of civilization, then, at the early stage of Chinese civilization, the dragon had already existed.

In the inscriptions of Oracle bones and bronzes, the word dragon is a kind of hieroglyph, which means the word of dragon followed the apparent shape of the dragon, and then was abstracted to obtain the perceptible character. However, there is not only one type of dragon; instead, according to historical records, there were more than seventy different characters for the dragon (Liu and Yang 1999). If the word dragon followed the apparent shape of something, usually animals, and there are so many words for the dragon, it means only one thing: the word dragon followed many real animals and possessed the common concept shared by those objects.

If we further inspect the Chinese words for dragon on the bronzes in the Shang Dynasty, even for the experts in this field, it is very hard to identify what animals are actually represented. Looking from those streaks on bronzes, how ancient people perceived the dragon was quite ambiguous and versatile. However, in the minds of the Shang people, those dragons had already conveyed certain common mysterious beliefs. From different kinds of bronzes, people prayed to dragons, inquired for happiness, and wished for hope, etc. Hence, no matter what the dragons really are, it had already evolved into a holistic concept related to the life and fate of people, possessing supernatural divinity, and had manifested itself through different animals.

In the Zhou Dynasty (1111–221 BC), the dynasty after the Shang, one of the earliest books in Chinese history, *Book of Songs* had already mentioned dragons many times: some of them were dragons on artifacts, but mostly they were dragons on flags. Most of the songs involving dragons were in "Soong" of the book. "Soong" collected the songs used by the Zhou emperor and the feudal lords in worship or important ceremonies. We can infer that only in the worshipping activities of the ruling class would there be flags with dragons on them. Therefore, at least in the Zhou Dynasty, the dragon had already carried the meaning of power and authority.

After the Zhou Dynasty, the Emperor Qin (221–206 BC) united China. The Zhou Dynasty was separated into two periods, Western and Eastern Zhou, because the capital was moved from the west to the east of China. The Eastern Zhou was also divided into the Spring and Autumn Period (722–403 BC) and the Warring State Period (403–221 BC) because the Zhou Dynasty was so weak it could not stop the feudal lords from fighting with each other. Qinshihuang was the first emperor to unite China. The Chinese name for Qinshihuang means the first emperor of Qin. Qinshihuang was also called the Patriarch Dragon, which means the first dragon of China. It was recorded formally in the first official history book of China: *Historical Memoirs*, compiled by Sima Qian of the Han Dynasty. From then on, the emperor is dragon and the dragon is emperor. The emperor was combined with the dragon and never divided until 1911.

The first emperor that actively created a legend of dragon for himself is the first emperor of the Han Dynasty, Liu Bang (254–195 BC) (Bo Yang 1977). After lasting only 15 years, the Qin Dynasty collapsed and was replaced by the Han under the leadership of Liu Bang. During that period, there were many warlords striving to conquer China. Since Liu Bang was too lowborn to compete with other revolution leaders, he had to create certain mythical stories to convince the uprising crowds to follow him. Because of this political necessity, he fabricated his divine birth by saying that his mother got pregnant by dreaming of the dragon. Further, every time he slept after being drunk, his friends always saw a dragon circling above him. This story would make people believe he was truly the dragon and destined to be emperor. Constructing this kind of fairytale is only to make people believe that he is the so-called "truly dragon son of heaven" and destined to be the emperor. No matter how much this tale contributed to Liu Bang's revolution, from the fact that this absurd tale was formally recorded in the first official history book, it is obvious that the Chinese people did believe the dragon is emperor and the emperor is dragon, or, at least that the emperor and dragon were inseparable.

At least after the Qin and Han Dynasties, the significance of the dragon was already very clear because the dragon had become the unique symbol representing the ruling power of the emperor. But, what exactly is the dragon, after all? Where does the dragon come from? It is a long-lasting myth, unsolved even today.

The concept of dragon is full of mysteries. There has never been an animal on earth called the dragon. The dragon is an artifact created by the Chinese. As long as it is an artifact, there must be certain values associated with the dragon. Since the dragon is created to be divine, it should be related to the worship of the ancient Chinese people. In ancient times, nature is full of terribly frightening forces, e.g., the wind, rain, thunder, floods, volcanoes, etc. Without any comprehension of these natural phenomena, spirits were created behind almost everything, a spiritual orientation of animism. The sun, moon, stars, mountains, trees, even animals, all became gods to be worshipped. The corresponding rituals to express and relieve human fear were also formed. The primary religion was created.

However, these gods must have some kind of symbols for people to worship. All kinds of different images of gods were then produced. These images should come from real things because they cannot be created from nothing. In certain periods of human history, there was an era of animism (Tylor 1871). At that time, humans began to recognize the superior capabilities of some animals and hoped to have certain relations with them. The images people worshipped were from certain animals or the combination of them.

In the prehistoric time of China, a legend said that the common ancestors of all the Chinese people Huang and Yen fought a war 5,000 years ago. Huang and Yen were leaders of their tribes. Huang defeated Yen with the help of animals such as tigers, bears, and elephants, together with some strange beasts. Obviously, those animals represented the tribes. Gradually, the capability, wisdom, and skill of humans grew. Animism became more complex and abstract. Compared with humans, certain animals were truly superior, e.g., very distinguished sensory capabilities, extremely strong, unimaginably long life, etc. Some animals can even fly. Certain characteristics of animals were admired by humans. Ancient people hoped to have the same remarkable competence, or at least be blessed or guarded. More abstract images were then created to reveal the expectations. It is believed that the concept of the dragon was formed during this period.

In the Shang Dynasty, the word dragon came from those images related to certain animals. There are many theories that attempt to portray the origin of the dragon (Liu and Yang 1999). Unfortunately, there is still no well accepted conclusion. Only recently as a result of several archeological findings, has one theory become gradually accepted by more and more scholars (Liu and Yang, 1999). The theory says that the dragon has multiple origins, that it evolved from six different kinds of actual animals. The evolution process started about 6,000 years ago and matured about 3,000 years ago. There were several animal images that had the characteristics of a dragon. Scholars then finalized six evident cases of these images to be the origins of the dragon. They are the oldest existing original images of dragons. They are called the protodragons. Each of these six different kinds of protodragon was created and used by the ancient people living in different regions of northern and eastern China. The six protodragons are the crocodile, tiger, snake, fish, pig, and hynobius. They are described as follows.

1. Crocodile

Close to the Yellow River in northern China, 6,400 years ago, there was a culture called Hougang, a type of Yangshao culture. In 1987, an ancient tomb was excavated (Pu-Yang Artifacts Management Council 1988). It is called the site of the West Bank of Pu-Yang, Henan. There were many potteries at this site with simple, unpainted pots. Only a small portion of them had simple marks and streaks. The people living here didn't seem very interested in painting on pottery.

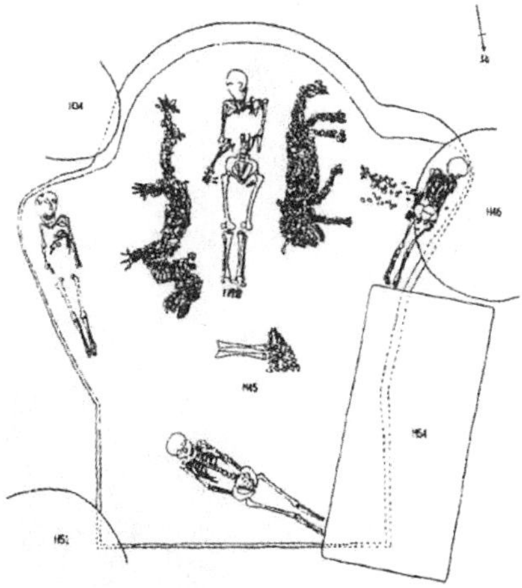

Figure 5.1 Crocodile protodragon – the first Chinese dragon, found in the West Bank of Pu-Yang, dated 6,400 years ago

Instead, they were very good at using clam shells to make vivid and lively figures. It is in the tomb numbered M45 that the so-called first Chinese dragon was found. In this tomb, two figures composed of white shells were found next to a skeleton at the center, as shown in Figure 5.1. To the east of the occupant of the tomb, it is obviously a dragon as seen from today's point of view. This dragon was made of carefully placed shells – with a length of 1.78 meters and a width of 0.67 meter – with its head pointing north and back to the west. The dragon, with a wide open mouth and big round eyes, had a long body and tail. There were two short legs, each with five claws pointing forward.

On the other side of the skeleton there was a figure composed of shells symmetrical to the dragon. It was obviously a tiger. There were other skeletons which were very likely human sacrifices. For these two funerary shell figures, one was a real existing animal, the tiger, and the other one was an unknown imaginary dragon. It seems illogical and senseless. It is well accepted that the dragon must also be a real existing animal. After a long debate involving considerable biological and historical evidence, most scholars gradually came to the conclusion that this figure was a crocodile.

Among the animals known to those scholars, including extinct dinosaurs, only the crocodile was closest to the figure of the dragon. From the layout in the tomb, both the tiger and the crocodile were actually servants of their master. The archeological study showed that the master of the tomb was a powerful figure in his society. The tiger and crocodile were used as a symbol to frighten the people and show the power of their master. As the tiger and crocodile are both terrifying animals with the capacity to bring about human calamity, only their master can control them. They thus represented the messengers of the sacred power behind them.

2. Tiger

In 1936, a Neolithic culture was discovered at Liangzhu, Zhejiang Province, which was dated between 4,000 and 5,000 years ago. Liangzhu was an independently developed regional culture with agriculture as the main economic activity. The handicraft trade was believed to be well developed. Among several excavation sites, different kinds of jade ware were found, which included Pai, Huang, and Jue, a split penannular ring which had a much smaller inner circle. On these jades, people discovered certain patterns of unknown animals similar to those found on Shang jades. They were accepted as one type of protodragon. Among them, the most typical one was a face called the dragon head on the bracelet.

In a tomb dating back 5,000 years, a jade ring was found (Jer Jian Archeological Research Institute 1988). The carving on the surface showed four strange dragon faces, as shown in Figure 5.2. After years of research, compared with other jades in Liangzhu, it was found that the dragon head on the bracelet was the simplified form of the typical god-human with animal face mark. The meaning of this mark is about a godlike hero sitting on the back of a terrifying beast. The beast, believed to be the dragon, had big round eyes and a big mouth with large teeth. Also, there were several stripes on the head of the beast and whiskers on both sides of the mouth. According to the Chinese tradition, the image is a typical resemblance to the face of a tiger. Therefore, the protodragon of Liangzhu was a tiger.

Figure 5.2 Tiger protodragon on the dragon head bracelet jade found in the Liangzhu culture, dated 5,000 years ago

The god-human with animal face mark was recognized as a combination of a man on top of a tiger. The religious implication of this mark was found to be a priest riding on a tiger and playing the role of a messenger, the medium between humanity and heaven. In ancient times, priests possessed the top position in a society. Through heaven, they gained powerful positions. The tiger was only a tool to symbolize their inscrutable power. Even the tiger bowed to them. Because the tiger is also a kind of terrifying animal, the main goal of using the tiger was to ask for obedience by frightening people with inscrutable power. The role of the tiger here was very similar to that of the crocodile mentioned previously.

3. Snake

Four thousand years ago, the Longshan culture also initiated one of the protodragons. The Longshan culture was in the late Neolithic age and was found in Shandong Province along the Fan River. Agriculture was already very prevalent. On a painted pottery dish found in an ancient tomb (Chinese Social Science College 1983), a strange figure with a piece of rice in its mouth was clearly shown on the bottom, as seen in Figure 5.3. The pottery dish was black with a red ring on the side. The curved creature was lying on the bottom. It seemed that this creature had two bodies, but only one head and one tail, although they were not quite clear. The eye was very small with a round shape. Sharp teeth were very evident. It was recognized as one of the protodragons of Shang and was called the spiraling dragon.

The spiraling dragon had no feet, and its head, neck, body, and tail could not be clearly discriminated. It was not any animal like a crocodile or tiger and others.

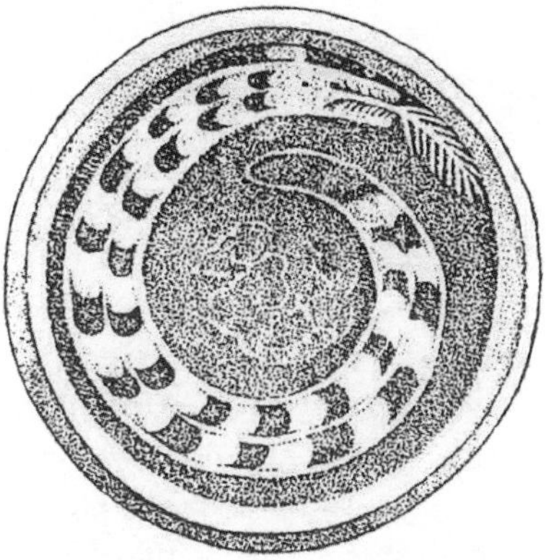

Figure 5.3 Snake protodragon on the painted pottery dish found in the Longshan culture, dated 4,000 years ago

It was obviously a snake. Together with many other artifacts found in this tomb, this dish was believed to be used for a cult, showing respect to the gods. The snake was used as a messenger for communication with the heavenly gods. Being also a terrible animal, the snake again acted as the symbol of the frightening inscrutable power behind it.

4. Fish

About 6,000 years ago, along the Wei River in northern China, an ancient group of people lived mostly by fishing. They represented one branch of the culture called Yangshao. Yangshao was a Neolithic culture discovered in 1921 at Yangshao Village, Henan Province. The most distinct feature of Yangshao is the large amount of painted pottery excavated. The site in Banpo, Xi An, is the most important one, although there are many different sites spread in a large area. The culture found is known as the Banpo type of the Yangshao culture. The people of Banpo already had agriculture, but hunting and fishing were still very important for their living. Most of the archeologists agreed that their society was matrilineal. The potteries found in Banpo were painted with different kinds of animals, such as goats, deer, and fish, among others. The largest amount of these potteries was painted with a great variety of fish. In addition to this, there were also many delicate artifacts made of fish bones, such as needles and necklaces, found in that site. It is quite clear that the fish was very important to the people of Banpo. Fish was not only food to eat but also something sentimental, even possibly something to worship.

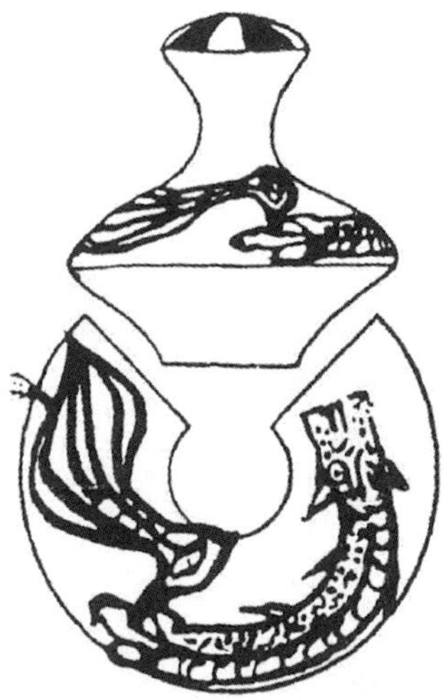

Figure 5.4 Fish protodragon on the bottle found along the Wei River, dated 6,000 years ago

A painted pottery bottle used for worship was found on the archeological site of one ancient village near Bao-Ji, Bei-Shau Hill (Chinese Social Science College 1983) of the Yangshao culture. The bottle was carbon-dated to be about 6,000–6,800 years old. A painting was clearly shown on the surface of the bottle. It can be recognized that the painting shows a bird catching a fish of a strange shape, as shown in Figure 5.4. The fish was recognized as one of the protodragons since its shape was not that obvious to be recognized as a fish. This "dragon" with a square head, round eyes and having several fins was painted on the shoulder of the bottle. A series of u-shaped curves appeared on the bottom of the "dragon." Behind the "dragon," there is a bird with its beak holding onto the tail of the "dragon." Archeologists concluded this painting to be a bird catching a fish. From the stripes around the eye of the bird, it was believed to be the cormorant. The ancient Chinese raised a special kind of bird and used them to catch fish in the river. Also, the grass ties clearly seen on the bird's neck were used to prevent them from eating the fish. This kind of fishing style still exists in different regions of China today. Hence, no doubt, the painting presented an image of a cormorant catching a fish. What did it imply? Obviously, the strange fish, which became a dragon afterwards, represents food, an essential resource of living. If a family can catch more fish, that means riches and wealth. Gradually, the fish or dragon acquired the implication of auspiciousness and wealth.

5. Pig

Five thousand years ago, in northeast China close to Inner Mongolia, there was a culture known as the Hongshan. The Hongshan represented a local culture in an area not too large. The remains of the Hongshan were found at different sites in the 1970s. The archeological findings included stone tools and some painted potteries. In addition, the most attractive findings in these sites were lots of well-made fine jades. There were all kinds of jades, such as: Bi, a flat disk having a circular concentric orifice in the center; Huan, a disk with a much larger orifice; Huang, a tablet in the form of half a Bi; Pei, a set of girdle ornaments worn at the waist; and many beads of different sizes. Those jades were believed to be used for ceremonial purposes or as decoration. In addition to these, there were many jades of animal shapes including that of a turtle, fish, bird, and tiger. Among the jades of animal shapes, the most attractive ones were the most difficult to identify which animal they really were.

In 1971, several ancient jades were found along the Liao River (Sun 1984). One of them, with a c-shape, was difficult to identify. The color of this object was dark green. The height was about 26 centimeters with a diameter of the cross-section ranging from 2.3 to 2.9 centimeters. There was a relatively long head on one side with a cocked mouth and flat nose. The eyes were long and narrow. The mouth was closed to form a thin line. On the back, there was long hair, like that on a horse, and with the end curved up. It is believed by most archeologists to resemble a pig, although there were still doubts and debates about what animal it really was.

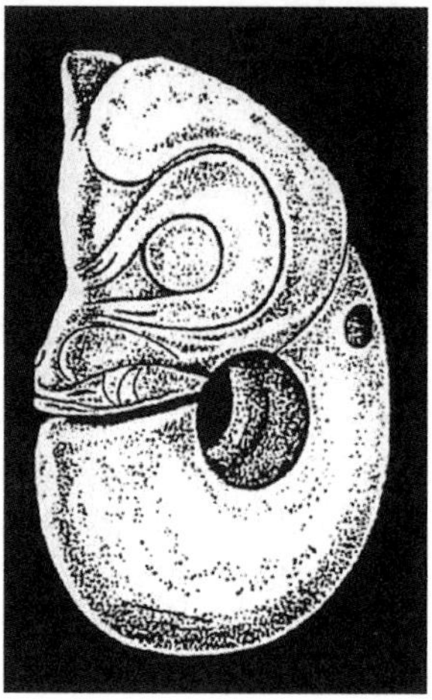

Figure 5.5 Pig-dragon jade as the protodragon found in the Hongshan culture, dated 5,000 years ago

In 1988, a second, much smaller, jade with a similar shape was found on the other site. This one was dated to be a little bit older. The c-configuration was basically the same as the previous one. There was one hole on the back for both of them and this proved to be used for hanging up. The curved portion was not the body of the animal but only represented a certain configuration. If the curved portion was neglected with the head left only, it would not be that difficult to identify what it was. From the general characteristics of the head, including the nose, mouth, eyes, and ears, most people will agree that this animal was a pig, as shown in Figure 5.5. Hence, these two strange pig jades were perceived as the dragon and are called pig-dragon jades today.

In the tomb where this pig-dragon jade was found, many pig bones were also excavated. At that time, farming crops and feeding pigs had already been adopted. Pigs represented the stable source of food. If a family bred many pigs, the family must be very rich because they had plenty of food. Pigs depicted wealth. Rich people then wore pig jade as a decoration to symbolize wealth, just as people of today wear diamonds. In addition, pigs were also used as sacrifices offered to gods in worship, a custom which still exists today throughout China, including Taiwan. Hence, the dragon gradually became the messenger carrying messages to the gods and bringing wealth to us.

6. Hynobius

In 1979 in the Gansu province, at a site representing a branch of the Yangshao culture –the Miaodigou – several pottery bottles dating back 5,500 years ago were found (Gansu Museum 1979). Miaodigou type was a culture parallel to the Banpo type.

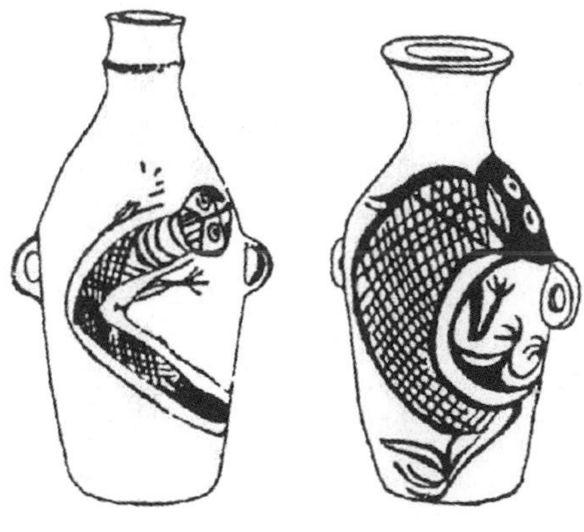

Figure 5.6 Hynobius protodragon on the painted pottery bottle found in the Yangshao culture, Miaodigou type, dated 5,500 years ago

The potteries of Miaodigou were mostly with black painting on a red base. The marks included numerous plants and flowers together with some fish, birds, and frogs. Among them, several strange paintings, as shown in Figure 5.6, showed an animal that attracted the attention of archeologists. The creature seemed to have a human face, round head, and big eyes, and the total height was 38.4 centimeters. It had a big mouth with evident teeth. Net marks were on the bodies. On the sides of the body, there were even two hands with four fingers each. The tail was so strange that it was not easy to identify what it was. Hence, this strange creature was also identified by archeologists as one of the protodragons.

After detailed study, it was recognized as the hynobius. And, this protodragon was then called hynobius with a human face. The hynobius is a small family of primitive Asiatic salamanders. Its Chinese name is "baby fish" because its head resembles a baby's head and its sound is just like the crying of a baby. Ancient Chinese regarded baby fish mysteriously as the human ancestor and created the concept of the family clan. It is believed that there must be activities related to ancestor worship using the baby fish as the totem. Gradually, after the baby fish was absorbed into the dragon, the dragon carried the implication of the ancestor of Chinese people. As portrayed in a famous pop song, which almost every Chinese can sing, "The Descendants of the Dragon," the dragon became the mythical ancestor of all the Chinese.

All these protodragons were obviously real animals, but, they also clearly carried very special significance and expectation. From this history, it can be seen that these six protodragons were developed independently in four different cultural regions of China. The pig protodragon was developed in the Yen-Liao cultural region, the tiger protodragon was developed in the Jiang-Jer cultural region, the hynobius protodragon was developed in the Gan-Qing cultural region, and the other three protodragons – fish, crocodile, and snake – were developed in the central China cultural region. No evidence of connection between them has been found. For thousands of years before Shang, these protodragons developed in parallel in the vast land of great China in their own separate environments, having their own characteristics, and without any blending or assimilating.

About 3,500 years ago, the Shang dynasty was established and ruled a relatively large area. The Shang territory covered the area from east to Shangdon, west to Shaanxi, north to Hebei, and south down to the Yangtze River. Those separate cultural regions then had a chance to merge under Shang rule. The prehistoric culture of China had multiple origins, as did the dragon. Following the merger of regional culture into Shang, the merger of protodragons into the unified dragon was also under way in the Shang period.

Since Shang ruled a large area, governing tribes with different beliefs and totems, it was then necessary to have a unified official ideology to rule effectively and efficiently. According to the historical records, the Shang people were extremely religious. They had so many gods to worship, with uncountable diverse images and rituals so complicated to an unprecedented level. However, from the excavated Oracle bones until now, among 150,000 pieces, there did not exist a unique supreme god or anything similar. The hierarchy of those spirits or supernatural entities was also not clear. That means the Shang people had many gods necessary to worship. Because of the demand from politics and religious belief, the ruling class of the Shang not only worshipped their own gods and ancestors, but also worshipped those from the others, even from different tribes.

Because of this, the ruling class of the Shang did not repel any of the different divine protodragons from other tribes; instead, they actively collected, accepted, and further inscribed those protodragons on different kinds of ritual appliances, and made them all serve the Shang emperor. This pattern, called "syncretism," is found in many ancient civilizations. These different divine figures, using ritual appliances as carriers, appeared in most of the Shang official ceremonial activities. The religious need of the Shang people was not only satisfied, but the other tribes were also content with such proclamation to some extent, therefore being appeased. Therefore, the convergence of the regional cultures into the Shang supplied the fertile soil for the merger of multiple protodragons. The inclusive political need and open religious attitude of the Shang helped the fusing process to finally shape the configuration and meaning of the Chinese dragon.

In addition to the characteristics from the six protodragons, there was one distinct feature of the Chinese dragon created in the Shang Dynasty: antlers on the head. From the type or pattern of the antlers, it looks very similar to one special character from Oracle bones. This word, as generally agreed by most of archeologists, obviously resembled the shape of the male genital organ. Putting male genital organs in the totem represented a concept of reproductive worship. Reproductive worship was very popular in ancient China, symbolizing the desire and expectation about life. Further, this character is also the ancient form of the modern Chinese word for ancestor. Casting a symbol of ancestor on dragon, obviously, it means ancestor worship. Therefore, the antlers on the head of the dragon imply expectation about life and worship to ancestor. Combining all of these, the Shang ruling of different tribes was believed to be quite effective by creating the idea of the common ancestor to acquire recognition and obedience.

In summary, these six protodragons and the antlers all depicted real animals and possessed clear implications. As the Shang people were very religious, as can be seen from the excavated Oracle bones, the ruling class of the Shang collected all these protodragons and absorbed them together. The ruling class did this in order to strengthen their power by blocking the communication route between the people and the gods, and creating the concept of common ancestor. From the sculptures portraying the dragon on the Oracle bones, it is believed that the dragon finally evolved during the Shang dynasty into a shape similar to that of today. Since then, only minor modifications have taken place.

From the evolution history of these protodragons and antlers, it can be found that the Chinese dragon possesses three attributes – because those protodragons can be separated into three different kinds of animal. Among the six protodragons, the crocodile, tiger, and snake are all terrifying animals. The cold-blooded crocodile lives close to the water. When hunting, it can silently approach the target with only its eyes above the water, or even totally submerge in the water without being seen. When close enough to be within strike range, the crocodile can attack with lightning speed. With such highly evolved killing skill, together with being located at the top of the food chain, the crocodile naturally became the frightening animal to the ancient Chinese.

On the other hand, the same role as the crocodile of king of the animals on the ground is the tiger. The tiger is also a kind of terrifying predator. In different human cultures, the tiger always possesses the characteristics of strong, ferocious, agile, and fast. Hence, it is reasonable that ancient people were fearful of the tiger. Since the tiger was one of the protodragons, the Chinese dragon then inherited the implication of a tiger. There is a long-lasting Chinese saying about the tiger and the emperor: accompanying with the emperor is accompanying with a tiger.

As for the snake, it is also a fearful animal. Although the snake is not as strong as the crocodile and tiger, it is elusive and hard to detect and defend against, and sometimes is even more frightening. In addition, the snake has a weapon which the crocodile and tiger do not – its poison. A single fast bite can cause death. All of these together make the snake a gruesome animal, like a ghost. Almost in every culture, the snake is a symbol of frightening evil, and in many cultures there exists a collective unconsciousness of the snake. In Christianity, the snake is even the incarnation of Satan. The snake, crocodile, and tiger are all predatory animals located near the top of the food chain. They all possess the power which human cannot reach, cannot understand, cannot predict, and cannot control. Carrying these images as the protodragons, the Chinese dragon thus partly represents frightening and inscrutable power.

As for the second kind of protodragons, the fish and pig were both used to represent wealth and auspiciousness. When the ancient Chinese knew how to breed the plants and animals, the sources of food became stable and the risk of not surviving dropped. Living standards were greatly improved. After raising the cormorants to catch fish, more fish represented more food, hence the more affluent and prosperous. With this, the meaning of fish would naturally be much more than just a fish. Painted on the pottery, used to pray or give thanks for a bumper harvest of food, the meaning of fish also expresses the expectation of good living – likely a very convincing idea.

As for the pig, the meaning was even more evident. In Chinese culture, the image of the pig traditionally means blessing. The pig is extremely useful in Chinese food since everything of the pig can be cooked into the dish. It clearly represents that the Chinese have heavily relied on the pig as the main source of food for thousands of years. Further, the Chinese character "home" portrays the image of a house with a pig inside, which means that only when there is a pig in the house can the house be called a home. When a "home" has more pigs, it must be richer. Consequently, pigs symbolize wealth, riches, affluence, and prosperity to the Chinese. These implications of fish and pig protodragons were passed down the generations through Chinese history to today. Hence, the Chinese dragon signifies the good and auspicious things one can expect. It can also be described as the auspicious expectation.

The hynobius and the antlers on the head imply the expectation of life and the ancestors' worship to the ancient Chinese people. It delineated that all the people living in different parts of China were actually the descendants of the same ancestor, implying that we all have the same ancestor and we are all in one family. Hence, with the hynobius and the antlers as the protodragons, the Chinese dragon naturally bears the meaning of family clan, and the family clan then became the third attribute of the dragon. This mentality, further strengthened by Confucianism, resulted in paternalism prevailing over every corner of Chinese society. The dragon further symbolized the core of the patriarchy, or simply the big father. It further solidified the authoritarianism of China.

Hence, the dragon as the unique symbol of traditional Chinese authoritarianism thus conveyed at least the following three essential ingredients:

1. inscrutable power;
2. auspicious expectation;
3. family clan.

Bearing these implications, the dragon was naturally used by the emperors to imply their frightening and beneficial power. Within 3,000 years, the dragon has permeated into the mind of every single Chinese. After the first emperor of the Han Dynasty claimed that he was the direct son of the heavenly dragon, the dragon became the formal symbol of Chinese emperors. Thereafter, almost every emperor used the dragon as the symbol of his power, except when the emperor was a woman. There were certain periods in which the dragon was even strictly prohibited to be used by common people (Liu and Yang 1999). It could be used only by the imperial family. From the numerous different types of dragon in every corner of the Forbidden City in Beijing, it is very easy to realize that the dragon has been the mature symbol of Chinese authoritarianism for a long period of time up to the present day.

Being the offspring of the dragon, every Chinese man and woman bears more or less some of what we might call "dragonality" or "dragon-ness." If one is the head of a group, whether the group is a family, a team, a class, a company, a school, a city, a party or a country, one has to rule like a father, educating every member that we are all in one family, asking obedience, showing authority, bringing the subordinates wealth and happiness. On the other hand, if one is a subordinate in a group, one has to show obedience and respect, following orders without critique or question, expecting to be taken care of. Hence, Chinese authoritarianism, signified by the dragon, represented a complete and rounded concept. Unlike the definitions of Western authoritarianism, including that of the Frankfurt School as represented by Theodor Adorno's *The Authoritarian Personality* or quantified by Hofstede, Hampden-Turner, and others, or "power distance," Chinese authoritarianism manifested a balance between payments and receipts, or expenditure and revenue for both the superior and subordinate.

Chapter 6
Measuring Chinese Authoritarianism in the Cockpit

In 2000, the Institute of Aeronautics and Astronautics, National Cheng Kung University, formed a research team to execute a study which was funded by the Aviation Safety Council of the Taiwanese Government. The project was called *Human Factors in Flight Safety – Study of Crew Resource Management and Corporate Safety Culture* (Lu et al. 2000). The goal of this study was to evaluate the status and effectiveness of crew resource management (CRM) training in Taiwan. This team was assisted by a team of anthropologists from Wayne State University in Detroit, including Dr. Allen W. Batteau, Capt. Alejandro Perez, and Ms. Carolyn Psenka.

Crew resource management refers to the better recognition and utilization of all available resources, information, equipment and people to achieve safe and efficient flight operations. As already known to the aviation community, many problems encountered by flight crews have very little to do with the technical aspects of flying. Instead, according to accident investigations, approximately 70% of air carrier incidents and accidents have been caused by the failure of flight crews to make use of readily available resources. Training pilots in a multi-crew environment to effectively use all available resources was then very important to further improve global flight safety.

Crew resource management (CRM) was developed in the US through cooperation between the National Aeronautics and Space Administration (NASA) and the University of Texas in 1980. Originally, it was designed for astronaut training, but was then used by the aviation community. It started out as cockpit resource management and then was expanded to include the entire flight crew, both on the flight deck and in the cabin. Starting from crew members in the cockpit, CRM gradually included the cabin members, air traffic controllers, and maintenance personnel. Now, it is referred to as corporate resource management to cover the whole company. With the United States still being the leader of global aviation, CRM was quickly adopted by airlines worldwide. Almost all airlines treat CRM as an effective tool to improve safety and promote corporate safety culture. Consequently, CRM has become a shorthand expression for global flight safety.

In 1990, China Airlines of Taiwan imported a CRM training program called cockpit resource management. The other airlines quickly followed. In 2000, the Civil Aeronautics Administration of Taiwan set up rules about CRM requiring airlines to comply – flight safety was already a critical issue in Taiwan after several major accidents. The core spirit of CRM as understood by the Taiwanese aviation community was synergy, i.e., marrying technical proficiency with effective crew coordination.

Following the original US design, CRM training should include six major areas:

1. **Communication and interpersonal skills**: The training is about certain special skills with good communication practices, including such items as polite assertiveness and participation, active listening and feedback. Of course, cultural influences must be taken into account to reduce the communication barriers induced from rank, age, and position.
2. **Situational awareness**: This refers to one's ability to accurately perceive what is going on in the cockpit and outside the aircraft. This awareness partly relies on cross-checking and ongoing questioning. Therefore, how pilots interact with each other would definitely influence performance.
3. **Problem-solving, decision-making, and judgment**: Basically, problem-solving can be seen as a cycle of events beginning with information input and ending with a final decision. In the cockpit, all decisions must come from the pilot-in-command and be supported by all crewmembers. The team will fail if command authority is not acknowledged at all times. Hence, the success of the decision also relies on the perception of authority and the flow of information.
4. **Leadership and followership**: In this portion, there should be clear recognition about the command role and interaction. The pilot-in-command is responsible for accessing and managing all resources available and the non-command crewmember is responsible for actively contributing to the team effort, monitoring changes, and being assertive when necessary. The primary base of the interaction between leader and follower is power. How people perceive power operation will play a critical role and have a definite influence.
5. **Stress management**: Any kind of emergency situation generates stress. Skills related to stress management refer not only to one's ability to anticipate, recognize, and cope with one's own stress, but also to perceive and accommodate the stress in others. That is to say that the crewmember should be on alert for any performance decline or subtle incapacitation. Certainly, this would include stress related to interpersonal problems.
6. **Critique**: The goal of critique is not to dwell on the negative, but to accentuate the positive and to encourage participation from the team. No doubt, the success of this training does depend highly on the general perception about accountability and the interpersonal relationship.

Crew resource management was invented by Americans. As an American artifact, it definitely bears the imprint of a Western value system. It has been widely accepted internationally that Hofstede's four-dimensional value system is the most useful framework to explain the key differences between national cultures

(Hofstede 1980), although its limitations for explaining Chinese authoritarianism have already been noted. These are power distance, individualism, uncertainty avoidance, and masculinity. Among these, the main scales which Western people use to measure the differences between national cultures are power distance and individualism. It could be further confirmed from the results of Helmreich (Helmreich and Merritt 1998). The other two were not that important since no big differences were found for different countries. Power distance and individualism are typical scales produced by Western linear thinking. From the contents of CRM, we can see the traces of power distance and individualism everywhere. Further, low power distance and high individualism are naturally assumed.

With these values, Americans designed crew resource management, assuming low power distance and high individualism. When such a highly Westernized artifact was imported into the sphere of Chinese culture, redefinition was inevitable. Take leadership as an example: when a Chinese pilot sees the word "leader," if it is in Chinese, the image surfaced in his brain would be largely different from that in an American's brain. Based on this, the activities of CRM such as communication, decision-making, and leadership, etc., in a Chinese cockpit would surely not be the same. Subsequently, after CRM was imported into Taiwan, if a detailed study about the effectiveness of this training was undertaken to find the gap between the original design and the implementation, as well as to further reveal the influence of culture, it would be extremely meaningful and important.

The study was called *Human Factors in Flight Safety – Study of Crew Resource Management and Corporate Safety Culture*. An unprecedented large-scale survey with a modified questionnaire was executed for this study in Taiwan. In many countries, CRM was an import for many airlines. No matter how CRM was executed, it was respected as a useful tool. After CRM had been adopted in Taiwan for many years, the training, management, experience, effect, and difficulties were all to be evaluated. Hence, the original questionnaire, Flight Management Attitudes Questionnaire (Helmreich and Merritt 1998), was used to reveal the effect of original design. Also, it is agreed that CRM has a built-in cultural bias. Thus, how the Taiwanese pilots perceived and reacted was to be investigated. So, the Flight Management Attitudes Questionnaire was modified to a certain extent to portray the Chinese influence on CRM in Taiwan.

Before designing the questionnaire, all the factors which could possibly influence the CRM performance were taken into account. Meetings with local airlines about the problems encountered when practicing the CRM training were called with the help of the Aviation Safety Council and the local Airline Pilots Association. Interviews with the management, flight operation and training departments and discussions with ten pilots from three local airlines were executed to help the preliminary design. In order to avoid bias and to have a more objective questionnaire, 22 pilots from different airlines were chosen randomly to run a pre-test and asked for their response. After collecting the corresponding comments, and finishing the first modification, a re-test done by 50 pilots was performed to finalize the questionnaire.

The questionnaires were grouped into three categories: CRM concept, CRM commitment, and CRM skill. Each category was divided into six subcategories:

1. communication;
2. situational awareness;
3. teamwork;
4. workload management;
5. decision-making;
6. culture.

All of these formed a three-by-six matrix. Basically, these questionnaires were designed to measure how the above-mentioned items were executed. A total of 58 variables were designed to evaluate the CRM performance, which included perceived management attitude toward CRM, stress understanding, open communication, non-punitive atmosphere, leadership, power distance, company training, cultural barriers in training, Standard Operating Procedures, safety reporting audit system, language skill problems, communication style, hiring standard, promotion standard, safety planning, company value employees, loyalty, working environment, flight operation satisfaction, perceived safety record, etc. There were 88 questions developed from those variables. Most of the questions used a five-point Likert scale, representing strongly agree, agree, neutral, disagree, strongly disagree, or very frequently, frequently, sometimes, seldom, very seldom. From 5 points to 1 point were assigned to each answer (5 = "strongly agree", 4 = "somewhat agree" … 1 = "strongly disagree"). The details of this questionnaire can be found in the report (Lu et al. 2000). Altogether, 1,947 copies were distributed throughout the management and pilots in six major airlines in Taiwan with the help of the aviation authority of Taiwan. Only 1,064 were effective, although in survey research this 55% response rate would be considered reasonably good. From the data, some very meaningful results were found.

In this study, how Chinese culture influences the effectiveness of CRM was also one of the main topics; the emphasis was then placed on the fundamental differences between Chinese and foreign pilots. Because of the policy of local companies, very few foreign first officers were hired to work in Taiwan. Consequently, the comparisons were made solely between the local and foreign captains. Among the questions directly related to differences of nationality, the means and standard deviations were obtained. The student's t-test was also used to check the significance of the difference in the answers between the local and foreign captains. The significance level was chosen to be $\alpha = 0.05$ here. In the study population there were 361 local and 188 foreign captains. As for the foreign captains, they came from all over the world, including Asia, North America, South America, Europe, Africa, and Australia. Those countries were the U.S., Canada, Mexico, Brazil, Venezuela, El Salvador, UK, Germany, Switzerland, Italy, Ireland, Holland, Austria, Norway, Bulgaria, Indonesia,

Malaysia, the Philippines, Iran, Korea, Australia, New Zealand, and South Africa. The six countries with the largest pools of pilots were the U.S., Canada, Venezuela, Mexico, the Philippines, and Indonesia. Although the foreign pilots do not represent a random sample of world cultures, one can accept that the 188 foreign captains can establish a contrast between Chinese culture and world cultures, since the pool of foreign captains covered a large enough area and the sampling was reasonably random.

As known to the aviation community, the attitude toward authority had certain degrees of correlation with global flight safety (Jing et al. 2001). On the other hand, the research results of Helmreich also proved that the strongest differences between pilots of different countries were observed with regard to the attitudes toward the personal, legal, and technical authorities (Helmreich and Merritt 1998). Hence, how Chinese authoritarianism interacts with the Western-designed CRM was then a critical issue in this study.

From the evolution of the Chinese dragon as described in Chapter 5, it can be found that traditional Chinese authoritarianism possesses at least the following three essential ingredients of the dragon:

1. inscrutable power;
2. auspicious expectation;
3. family clan.

Being the offspring of the dragon, every Chinese person, including pilots, bears more or less some dragonality. In the cockpit, the captain and the first officer form a group with a very clear power structure. The captain is expected to be the leader while the first officer is the follower. From the previous discussions about the dragon, dragonality would be expected to play an essential role in the Chinese cockpit. From the data obtained in the CRM study, the main comparison was made between the local and foreign captains. From certain questions with a statistically significant difference, Chinese authoritarianism could be clearly depicted. According to the dragonality discussed above, the representative questions were separated into two groups here to portray the characteristics of Chinese authoritarianism. The first is inscrutable power and the second is auspicious expectation. As for the third trait of dragonality, family clan, the contents of it include using the family as the model of hierarchy projected onto any group, maintaining harmony within the family, and protecting the interests, honor, and even face of the family members. However, to clearly depict this, the present CRM survey was not sufficient. It would need deeper research to explore the real characters related to family clan, because the concept of family in China is even more fundamental than authoritarianism. Consequently, it is omitted presently and will be discussed briefly in the next chapters.

Table 6.1 Questions related to inscrutable power

N	Question	Taiwanese CAP Mean	Taiwanese CAP Standard deviation	Foreign CAP Mean	Foreign CAP Standard deviation	t-test t_{crit} = 1.645 $\alpha = 0.05$
Q24	I am reluctant to express disagreement with my superiors.	3.10	0.99	2.37	1.11	7.87
Q59	When I am communicating to my superiors, I need to be careful how I approach him.	3.64	0.85	3.19	1.17	5.26
Q1	If I perceive a problem with the flight, I will speak up, regardless of who may be affected.	4.40	0.80	4.65	0.66	3.67
Q16	During an emergency I still like to ask the crewmembers' opinion before I make a decision.	3.74	0.98	4.01	0.96	3.01
Q43	I feel comfortable going to my manager's office to discuss problems or operational issues.	3.21	1.10	3.48	1.19	2.61

6.1 Inscrutable Power

In the present questionnaire, among the questions in which the differences between Taiwanese and foreign captains were statistically significant, i.e., t greater than t_{crit} for $\alpha = 0.05$, there were 13 questions related to superiors and subordinates. In the first group, there were five questions about the inscrutable nature of power. The questions, together with means, standard deviations, and t-values, are all listed in Table 6.1. In this table, the questions are listed according to the order of t-values, with highest t on the top.

In these five questions, the very first question with a t-value of 7.87, is question 24. This question asked:

> 24. I am reluctant to express disagreement with my superiors. (5 = "strongly agree" … 1 = "strongly disagree")

This question represented the biggest difference in results between the local and foreign captains. The average of Taiwanese captains is 3.10 and the average for foreign captains is 2.37. The data showed that the Taiwanese captains were much more reluctant to express disagreement to the superior compared with foreign captains. That is to say, the most distinctive character of Chinese people, at least in the cockpit, when facing superiors is obedience, i.e., they would never want to say no to superiors, even when they did not agree with them. A possible explanation is that the subordinates don't know what the superior is going to do after they have heard an opinion against theirs. The worry comes from the unclear intention of the superior's power. In this case, the risk to confront a superior would be higher. This result was thought to be consistent with the general image of the Chinese. This is also the most typical feature of traditional Chinese authoritarianism: obedience.

Question 59 had the second highest t-value, 5.26. It asked:

> 59. When I am communicating to my superior, I need to be careful how I approach him. (5 = "very frequently" … 1 = "very seldom")

The average is 3.64 for the local captains and 3.19 for the foreign captains. From the data, it is obvious that the Taiwanese captains were really much more careful, even including the words they used, when they were facing their superior. That meant that the Chinese subordinates felt that they had to be very careful when dealing with their superiors, i.e., be sure to show respect, not to express any disobedience, in order not to provoke their superiors. The explanation for this is relatively similar to the previous question: uncertainty about the intention of the superior power. From the results, it can be seen that the foreign captains would be less likely to think they should be very careful in dealing with their superiors. It is obvious that the power seems to be not that terrible to them.

The superiors mentioned in the above questions were the same group of leaders from different airlines in Taiwan. Those foreign captains were all working in Taiwan. They faced the same superiors as the local captains did. While the Taiwanese captains were reluctant to express disagreement and needed to be careful to approach superiors, foreign captains were not: one would be confused about whether the superiors were polite or harsh. When two different groups of captains, foreign and Taiwanese, were faced with the same superiors while having significantly different perceptions about the superiors, it is hard to believe that this difference would come from the superiors because the superiors were the same group of people. It is even not possible to say that there was discrimination between local and foreign captains. The difference could only be from the captains themselves. The same power was felt to be polite to foreign captains while harsh to the local captains. Hence, this difference had nothing to do with the power itself, but was created by the perceptions of the pilots. In addition, there were so many captains having the same perceptions about power that it can only be explained by culture – in this particular situation, by the traditional Chinese authoritarianism.

Question 1 had the next highest t-value, 3.67:

> 1. If I perceive a problem with the flight, I will speak up, regardless of who may be affected. (5 = "strongly agree" … 1 = "strongly disagree")

The average for Taiwanese captains was 4.40, while it was 4.65 for foreign captains. From the absolute value of the data, Taiwanese pilots still seemed likely to speak up when they perceived a problem. During flying, any problem would possibly evolve into a serious one if it was left uncorrected. Under normal conditions, every trained pilot would raise the issue and find a way to deal with it. However, compared with the foreign captains, it can be seen that fewer local captains will speak up, which means that more local captains hesitate to speak up even when perceiving a problem. To the Chinese, whether the action will affect somebody is really a serious concern influencing whether to speak up or not. If the other side affected was from a higher level, the problem will be even more serious because the consequence will be more risky. It is quite clear that this mentality is from fear of being punished. Even if the problem being raised would not affect anyone, if it was proved to be wrong, at the least it will mean losing face for the pilot who spoke up.

Question 16 in Table 6.1 has a t-value of 3.01.

> 16. During an emergency I still like to ask the crewmembers' opinion before I make a decision. (5 = "strongly agree" … 1 = "strongly disagree")

The average is 3.74 for local captains and 4.01 for the foreign pilots. The data revealed that during an emergency fewer Taiwanese captains liked to ask the opinion of crewmembers before making a decision. By definition, the captain is the one who should make the final decision and be responsible for the flight. In the Chinese mind, since I have the power, if I ask your opinion, it is my leniency. If I don't, you still should not question why I did not ask you. During an emergency, the subordinates should be even more reliant upon their superiors. Hence, whether it is the responsibility of the captain or reliance on the first officer, as the captains had the power to make final decisions, the first officer did not need to know or even speak up, especially in case of an emergency. The foreign captains apparently did not think that way. Chinese authoritarianism is more responsible in nature as the role of a father in a family. On the other hand, keeping the power obscure and inscrutable is the traditional Chinese way of exercising power.

Question 43 had a t-value of 2.61. It asked:

> 43. I feel comfortable going to my manager's office to discuss problems or operational issues. (5 = "very frequently" … 1 = "very seldom")

It was not surprising that the average of the local captains, 3.21, was lower than that of the foreign captains, 3.48. Taiwanese pilots would not feel comfortable going to see the superior and talking freely compared with the international pilots. Why is that? Traditional Chinese education teaches every generation, from when they were still children, to behave appropriately, especially when meeting with the elders. Every Chinese would then restrain himself/herself to show respect to the superior. It is also a kind of protection behavior: when the intention of the superior power is not clear, it is better to be careful. To the Chinese, the power is always inscrutable.

Grouping all of these data and discussion, it can be found that the Chinese have deeper fear when faced with the superior. The basic reason is that the superior power is usually more capable and influential, together with an unclear intention, just like the feeling when we suddenly find that there is a crocodile, tiger, or snake right in front of us. The data here clearly shows the perception of Chinese people regarding power or authority. This mentality is actually the essence of traditional Chinese authoritarianism. Also, this mentality is very close to the unconsciousness represented by the totem of crocodile, tiger, and snake in dragonality. Consequently, it could be reasonably concluded that the frightening inscrutable power was thought to be the major characteristic of Chinese authoritarianism.

6.2 Auspicious Expectation

On the other hand, there were eight questions in the second group related to superior and subordinate, with significant differences between the local and foreign captains. These questions were related to expectations about power. Stated simply, Western people also fear superior power; this may be universal to all humans. The difference is that in Chinese culture the cruel part of operation of power was much more extreme. However, the other side of Chinese authoritarianism – auspicious expectation – was very difficult to find in the perception of power in Western culture. In the Western definition of power, it is very hard to find any kind of special expectation. In China, because of the long history of the dragon culture, the Chinese people have a relatively high expectation about power. Just like the implications of the symbols of the fish and the pig in the dragon, the Chinese expect power to be as the fish and pig imply – to bring them identity, status, wealth, and dignity.

Inscrutable power and auspicious expectation were not mutually independent: they are two sides of the same coin. When a subordinate is reverent and respectful and also hears and obeys the superior, it is very likely this was not a casual attitude. They must be expecting something in return. On the other hand, in this culture people may be fairly good at figuring out the intentions of the superior. At least, they would be very careful about the intentions of the superior, so as to meet the superior's need. Thereafter, most likely, the superiors will find they were more useful and can fulfill their will compared with others. Consequently, this kind of subordinate will have more chance to get rewards and good positions. Hence, showing respect and obedience to the inscrutable power may not always be in vain. It might come with an expectation. Whether from the perspective of superior or subordinate, inscrutable power and auspicious expectation are truly twins.

The most fundamental reason for the Chinese that inscrutable power and auspicious expectation are truly two sides of the same coin is the traditional Chinese concept of "pao," which primarily means "in return" (Yang 1957). Pao is accepted as a basis of Chinese society. Pao is an identified interaction pattern of Chinese people between human and human, human and nature, and even human and super nature. Chinese people believe firmly in this kind of cause-and-effect relation traditionally. Concepts related to pao exist almost everywhere in Chinese society. The concept of pao is relatively similar to Newton's third law of motion: for every action there is an equal and opposite reaction, or in other words: where there is an action, there is always a reaction, but exceeding the limit of the physical world and extending into human society. Naturally, pao also exists between superior and subordinate.

In the second group of questions, one can see clearly the expectation side of Chinese power. Firstly, there are two reciprocal questions related to leadership style which obviously reveal the local captains did have higher expectations of their superiors and thus lower satisfaction. In Table 6.2, the captains were asked about leadership styles. Four styles were listed. Starting from a more autocratic style with A – where the leader usually makes decisions promptly, communicates them to subordinates clearly and firmly, and expects them to carry out the decisions loyally and without raising difficulties – to a more democratic style with D – where the leader usually calls a meeting of subordinates when there is an important decision to be made, the leader puts the problem before the group and invites discussion, and accepts the majority viewpoint as the decision. A range of 4 to 1 points were assigned to each answer.

For question, 66, with a t-value of 2.97, the pilots were asked:

66. Which one of the above styles of leadership would you most prefer to work under? (4 = "style A" … 1 = "style D")

This question was about the preference to work under a particular style of leadership. A lower mean value represents a preference for a more democratic style. It was found that local captains got lower scores, 1.85, while the foreign captains got 2.04 – the difference was statistically significant. It is clear that all the pilots preferred a democratic style of leadership; however, local captains preferred this more compared with foreign pilots. Obviously, the international captains seemed to not value egalitarian decision-making as much as Taiwanese pilots did.

Question 67 asked:

67. In your organization, which style do you find yourself most often working under? (4 = "style A" … 1 = "style D")

This question asked about the existing leadership style. It asked the captains to describe their working environment in Taiwan. It was found that the local captains often found themselves mostly working under more autocratic leaders, since the average score was 3.48. The average of the foreign captains was 3.17. And, the difference was even larger than that of question 66, with a t-value of 3.84. From the data, the Taiwanese captains thought that the leadership style in Taiwan was between styles A and B, while the international captains thought it was close to style B. Again, both the local and foreign pilots were talking about the same group of leaders.

Table 6.2 Questions and data related to leadership style

Please read the following descriptions of four different leadership styles, and answer the questions that follow:

Style A Leader usually makes decisions promptly and communicates them to subordinates clearly and firmly, and expects them to carry out the decisions loyally and without raising difficulties.

Style B Leader usually makes decisions promptly, but, before going ahead, tries to explain them fully to subordinates. The leader gives the reasons for the decisions and answers whatever questions they may have.

Style C Leader usually consults with subordinates before reaching decisions, and listens to their advice, considers it, and then announces the decision. The leader expects all to work loyally to implement it whether or not it is in accordance with the advice they gave.

Style D Leader usually calls a meeting of subordinates when there is an important decision to be made. The leader puts the problem before the group and invites discussion, and accepts the majority viewpoint as the decision.

		Taiwanese CAP		Foreign CAP		t-test
N	**Question**	**Mean**	**Standard deviation**	**Mean**	**Standard deviation**	$t_{crit} = 1.645$ $\alpha = 0.05$
Q66	Which one of the above styles of leadership would you *most prefer* to work under?	1.85	0.77	2.04	0.68	2.97
Q67	In your organization, which style do you find yourself most often working under?	3.48	0.84	3.17	0.99	3.84

The local pilots thought that the leaders in Taiwan were more or less autocratic and the international pilots did not, although they were all talking about the same group of Taiwanese leaders. We cannot expect that the same group of leaders would have two different leadership styles. The difference could only come from the perceptions of the pilots. Then the question would be why local and foreign pilots have different perceptions.

It is reasonable that higher expectation usually leads to lower satisfaction. It is quite obvious that the Taiwanese pilots had an over-expectation that the leadership style would be democratic. By looking at the question, they expected to be asked and consulted by the superior, and for their opinions to be respected and valued, while those foreign captains working in Taiwan did not. The higher expectation as shown in question 66 will certainly result in a higher degree of discontent, as shown in question 67. If my boss does not ask my opinion, he must be autocratic. From these data, it is quite clear that the Chinese people do have a higher expectation about the superior compared with the foreigners. It is not surprising, because the dragonality originally has two faces of authoritarianism, i.e., inscrutable power and auspicious expectation, as implied by the crocodile, tiger, and snake, and the fish and pig as protodragons.

The auspicious expectation of Chinese authoritarianism will be even more evident in the following questions. These questions also revealed what local captains expected from their superiors in addition to the leadership style. Looking for the questions with lower satisfaction, one can understand what they expected. There are six questions listed according to the order of t-value in Table 6.3. These questions were all about the captains' perceptions of their superiors or how their superiors treated them. Only the questions with significant difference compared with the foreign captains are listed in the order of t-values.

Table 6.3 The higher expectation about superiors frequently resulted in lower satisfaction

		Taiwanese CAP		Foreign CAP		t-test
N	Question	Mean	Standard deviation	Mean	Standard deviation	t_{crit} = 1.645 $\alpha = 0.05$
Q61	My superiors are courteous and polite when they are talking to me.	3.54	0.88	4.07	0.79	6.99
Q69	The resources that my company devotes to employees' training are adequate.	3.26	1.09	3.81	0.96	5.86
Q82	How does your company support you in your job?	2.96	0.74	3.22	0.78	3.82
Q49	My immediate superior participates in CRM.	3.49	1.03	3.77	1.07	2.99
Q32	I think my company has a long-term flight safety plan.	3.16	1.10	3.43	1.10	2.71
Q28	Management will compromise safety concerns for profitability or any other reason.	3.26	1.26	3.00	1.21	2.30

Question 61 has the highest t-value, 6.99. It is about the attitude of superiors.

61. My superiors are courteous and polite when they are talking to me.
    (5 = "very frequently" … 1 = "very seldom")

The average of local pilots was 3.54 while the average for foreign captains was 4.07. The difference was quite significant. Far fewer local captains agreed that their superiors were courteous and polite to them. To the same group of leaders, Taiwanese pilots had lower satisfaction about the attitude. That means the local captains had higher expectations about the attitude of their superiors while the foreign pilots did not. They expected their superiors to have kindness toward them.

The same treatment was different in the eyes of local and international captains: international pilots thought it was polite enough and the local pilots did not.

The question with the second highest t-value, 5.86, is question 69, which described how much resource the company devoted to training.

69. The resources that my company devotes to employees' training are adequate. (5 = "ample", 4 = "adequate", 3 = "inadequate", 2 = "barely enough", 1 = "pathetic")

The average score of local pilots was 3.26, close to "inadequate," while the average score of foreign captains was 3.81, close to "adequate." The difference was again quite significant. From the scores shown here, it was hard to tell whether the resources the company devoted to training were enough or not. But obviously, Taiwanese captains expected their company to devote more resources on them because the company should of course take care of their subordinates and provide support to them. If not, how can the company ask for the subordinates' obedience and loyalty? If the obedience and loyalty were already paid to the company, the subordinates would naturally expect something in return.

The question with next highest t-value was question 82. It was about the support from the company.

82. How does your company support you in your job? (4 = "sufficient", 3 = "adequate", 2 = "inadequate", 1 = "not at all")

There were four choices, ranging from "sufficient" to "adequate" and then to "not at all." Taiwanese captains got 2.96, while foreign pilots had 3.22, with a t-value of 3.82. From the data, local pilots thought the company provided support at a level that was a little bit less than adequate. On the other hand, the foreign captains thought the company provided support that was more than adequate. Again, the local captains had less satisfaction about the resources and support they got in their jobs from the company. Put simply, the reason for this is again that they had higher expectations about the company taking care of them. In Chinese minds, it is very natural that taking care is simply the payback from superiors when they have already shown obedience and respect.

Another question with a significant difference between the local and international pilots is question 49. It asked:

49. My immediate superior participates in CRM. (5 = "very frequently" … 1 = "very seldom")

In this question, the average answers of local pilots were 3.49, while the international pilots were 3.77, with a t-value of 2.99. Whether the same group of superiors from the airlines participated in CRM would be a simple fact. However, the local captains and international pilots still had different perceptions about the same reality. No doubt, it was the expectation that caused the difference. The Taiwanese pilots expected their superiors to participate in CRM more so than the foreign pilots. In Chinese minds, the superiors should be much more capable than the subordinates so as to ask for their obedience and respect. Thereafter, they had to actively participate in CRM to be wise enough to be qualified superiors.

Question 32 asked:

> 32. I think my company has a long-term flight safety plan. (5 = "strongly agree" … 1 = "strongly disagree")

In this question, the average answers of local pilots were 3.16, while the international pilots were 3.43, with a t-value of 2.71. Compared with the local pilots, more international captains agreed that the company has a long-term flight safety plan. Again, whether the company has a long-term flight safety plan is simply an empirical question, not a matter of perception. Yet, the perceptions were still different, as anticipated. The Taiwanese captains were really always more harsh to their superiors as well as the company because of the higher expectation.

The last question in this group, question 28, was about the company's flight safety policy.

> 28. Management will compromise safety concerns for profitability or any other reason. (5 = "strongly agree" … 1 = "strongly disagree")

The average answers of local pilots were 3.26, while the international pilots were 3.00, with a t-value of 2.30. The significant difference still existed. From the data, facing the same group of management, more local pilots believed that safety was compromised by profitability or some other things. The foreign captains were neutral on this issue. Due to the higher expectation, the Chinese usually expect their superiors to be capable, wise, kind, and able to take care of them. Of course, the superiors should know not to sacrifice safety for any reason. Higher expectation always leads to lower satisfaction. Frequently, discontent or even resentment would grow. In this case, the subordinate would be very picky and critical of the superior.

From the above data, it is quite obvious that the Taiwanese captains, in comparison with international captains, expect their superiors to be polite, supply more support, more training, more resources, and participate in CRM, and to plan for safety and not compromise. Basically, in the minds of local pilots, the superior should be almost all-mighty, but with good will. And, it is of course not possible to find that kind of superior. Hence, disappointment follows. On the other hand, it was hard to find similar expectations from the international pilots since all the data from them were significantly different. From the perspective of the dragon culture, it was actually very reasonable. Through thousands of years of dragonality tradition, the Chinese generate auspicious expectation toward power, which is usually possessed by the superior. The Chinese expect that the superiors, having inscrutable power, should perform very well. Otherwise, how can the superiors have the capability to take care of us? How can the superiors deserve our obedience and respect? Consequently, the opposite side of the frightening inscrutable power of Chinese authoritarianism is auspicious expectation, which should include at least kindness, taking care, and superb superiors.

It can be seen that the concept of authoritarianism in the Chinese mind represents a complete idea as it was clearly depicted in the cockpit. On the one hand, it carries the frightening inscrutable power to punish the subordinates and request for obedience and respect. On the other hand, it bears the capability to bring the subordinates wealth and happiness. To take care of the subordinates is the inevitable obligation of the superiors. At the least, the superiors should be kind.

The research performed by National Cheng Kung University was an unprecedented large-scale study about aviation in Taiwan. The results were believed to be representative. From the data, the faces of Chinese traditional authoritarianism in the cockpit were clearly revealed, and found to possess the key features of the culture of the dragon. Further, it obviously could not be explained by the simple, Western concept of power distance. The most distinct difference of the Chinese concept about power from the West is auspicious expectation. As explained, the concept of power distance was thought to originate from the linear mode of thinking of Western culture. It comes from the gap, or distance, between the powers of two parties. Distance is the measure of the length of a line between two points. Therefore, using power distance to explain the operation of power would naturally emphasize the direct comparison of the power. This was believed to be the reason for the British envoy choosing not to kowtow while the Dutch envoy did.

On the other hand, traditional Chinese authoritarianism was derived from the holistic perspective about the power operation, based on the Chinese ideographic mode of thinking. The main concern of Chinese power operation was placed on the total effect, including the feedback loop between both parties. Power is not a uni-directional concept. It is understood and condensed as the simple concept of "pao" by most Chinese people. That is the basic reason for the appearance of auspicious expectation. By contrast, "power distance" does not contain the feedback and interaction of mutual cause and effect. There will then not be anything like auspicious expectation in the cockpit. In addition, with the more fundamental concept of individualism, Western people tend to stress self-reliance, so as not to expect to get something from the power operation. Thereafter, Western pilots do not especially need to show obedience and respect to the superior.

The study about Chinese authoritarianism in the cockpit is of course not complete. There are still many aspects waiting to be explored. However, the Chinese dragon is already accepted as the most appropriate symbol of Chinese authoritarianism, as can be seen from the long history of China. The implication of the dragon does provide a good starting point to clearly show the faces of Chinese authoritarianism in the cockpit. With the survey data, the concept of "dragonality" is demonstrated to be appropriate to some extent for describing the differences between the Chinese and non-Chinese pilots. In terms of technology issues, in which aspect and to what extent the designed Western CRM was influenced by authoritarian mentality deserve further detailed study. In addition, to clearly define how much and what kind of authority the operators have or should have, or how to share the authority between the automation and the operators, or the level of crew coordination required, would be the next appropriate direction to proceed along when facing authoritarianism in the cockpit. To tame the Chinese dragon in the cockpit, the Western aviation system should also evolve to cope with it. It is a culture-technology co-evolution process. It needs cooperation between both parties.

Chapter 7

We Are All in One Family

As used for 2,000 years by Chinese emperors, the dragon is clearly the most appropriate symbol for Chinese authoritarianism. From archeological study, the implications of the dragon were clearly revealed. From the six protodragons, the original real animals evolved into the dragon together with the antlers on the head. This can be separated according to the characters and grouped as (1) crocodile, tiger, and snake, (2) fish and pig, and (3) hynobius and antlers. Each of these had their own implications. The first of these is inscrutable power, which is thought to include traits such as obedience, respect, and fear of punishment. The second one is auspicious expectation, including features such as kindness, taking care, and the superiority of a leader. The last characteristic is represented by the hynobius and antlers. The implication was very clear that all Chinese come from the same ancestor. Hence, we are all in one family. We are all brothers and sisters. From the ancient time, traditional Chinese authoritarianism was full of the meaning of family. Hence, although the emperor is the typical symbol of Chinese authoritarianism, its prototype is truly the father in the family. Patriarchy became the fundamental model of power operation at different levels of organization, including country. In such a culture, the leader in any organization with a clear structure of superior and subordinate will behave more or less following the role of a father. The others will then accept their roles just like family members. In Chinese society, family is everywhere.

Family is the basic constructing element of Chinese society, from ancient times until today, and will be the same in the future. Since being born, everybody was imparted with long-lasting, unchangeable relations with parents, brothers and sisters, and relatives. Throughout life, everybody has to live and develop within different relations in a family. In anthropology, family is the primary group; different kinds of relations from the family are primary relationships. Every relationship is attached to a corresponding value of right or wrong, good or bad. This corresponds to a system of ethics. Ethics originally started from the family for every human society.

However, in the history of Western civilization, after the rise of Christianity, family ethics has always been repressed, or at least not encouraged. Thus, the primary relations from family would not thrive as they did in Chinese society. According to the Bible, Jesus tried to suppress family ethics. As recorded in Luke 14:26, "If any one comes to me and does not hate his own father and mother and wife and children and brothers and sisters, yes, and even his own life, he cannot be my disciple." It was very reasonable. Christianity accepts God to be the Heavenly Father of everyone. Most of the time, emphasizing family ethics would very likely obstruct preaching of the gospel from Christ. Hence, the obstacle from family must be removed. Riding the political strength of the Roman Empire, the basic beliefs of Christianity spread throughout Europe and became the indispensable source of many value systems in Western civilization. Therefore, the weakness of family ethics was thus inevitable.

On the other hand, in the long history of China, family as the primary group has never been suppressed by any religion. Moreover, because of the concept of Mandate of Heaven in the Chou Dynasty (Hsu 1990), which was created for political reasons combined with morality, the ethics of family were strengthened and standardized as the model to other fields of society, including politics. In China, gradually, ethics coming from family extended far beyond family (Liang 1963). The primary relations combined with different moral systems became various types of ethics, with which every Chinese from birth had to comply. The most important among them were the famous Five Cardinal Virtues. In Chinese culture, there are many different kinds of interpersonal relationships. Among them, five were identified to be the most fundamental. They are father–son, ruler–subject, husband–wife, older brothers–younger brothers, and friends. There is also necessary morality associated with them to be the Five Cardinal Virtues, which are *affection* between father–son, *righteousness* between ruler–subject, *distinction* between husband–wife, *order* between brothers, and *sincerity* between friends (King 1992). The Five Cardinal Virtues are like the Ten Commandments in the Bible, except there is no God.

Since then, the power of family grew indefinitely, and became the primary shaping force of the behaviors of every family member, and even the behavior of every organizational member. During the period between being born, as a blank slate, to becoming an adult, with full capability to behave, the mind of every Chinese person was implanted with these principles. Family ethics was thus programmed into the mind of every Chinese. After they left family to enter into society and country, the coding in their brain cannot and will not be reprogrammed. Therefore, no matter what kind of situation is confronted, family ethics is always the model projected onto it. Family ethics combined with morality molded the Chinese culture to be the ethical one, described by a famous Chinese scholar as the "religion of the good citizenship" (Kou 1915). And, the family became the stronghold of Chinese ethical culture. Although individuals can learn multiple cultures, in times of stress they frequently revert to their earlier, more basic cultural training, as Karl Weick noted in his analysis of the Tenerife air disaster (Weick 1990).

All of China is a hierarchical organization of family, a big family composed of countless medium and small families. The emperor is the father of the country-family, and the people are brothers and sisters. Every person is the emperor or the people, father or son, husband or wife. Every person was assigned several inescapable roles from family to play, throughout their entire life. Every role comes with certain values to comply with, such as the Five Cardinal Virtues originated from family. In such a society, it is not possible to have the concept as individual entity.

Individualism would not have any chance to take root and grow. In the soil of this ethical culture, it is not possible for the flower of individualism to flourish.

In Chinese society, using family as the prototype of any organization, compared with Western society based on the individual, interpersonal relations would definitely be very different. An interpersonal relationship based on individualism is expected to be relatively simple. The dichotomous, linear mode of thinking will automatically separate group from individual, so that the so-called interpersonal relation becomes the relation between individual and the group. As for the Chinese, living in different kinds of families throughout one's entire life, one would certainly experience a wide range of interpersonal relations instead of only that between the individual and the group. Interpersonal relationships in Chinese society are much more complicated.

If we are going to choose a representative core feature to portray Western society, it would be individualism. From Roman times up until today, no matter in terms of ideology or daily life, the West has always swung back and forth between the individual and the group. Of course, it is also concrete evidence of the linear mode of thinking of the West.

Forming the concepts of individual and group, Christianity had a strong influence. Christianity accepts God to be the Father, everybody being brothers, creating an organization, the church, exceeding the boundaries of family and tribe, and breaking the barriers from family and class. In the early Church, Christians lived together and developed their communal life. Early Christian congregations provided members with a strong sense of community, with mutual religious and material support. It is understood that those who joined the earliest Christian communities were the poor and the marginalized. They had suffered much injustice at the hands of the religious and political authorities during their day. The Church thus provided the essentials of social security. The most obvious feature of their community life was that believers had everything in common; they sold their possessions and their goods and distributed these among all in accordance with each one's needs. This rejection of private property was a prerequisite for being a member of the community. Christians always thought in terms of the whole community. In this early faith community, as a consequence of the rejection of private property, it could very well be concluded that the social order was that of a classless society and in the kingdom of heaven there cannot be social differences. Karl Marx did not invent the idea of a classless society; it was Jesus Christ who originally taught it. And it was practiced by the early Christian community.

However, over many centuries, as the Christian faith was institutionalized in the Roman Catholic Church, the individualism of the early believers was subordinated to the hierarchy of the Church. The corruption of the Church led to the Reformation, an awakening of the self to follow its own conscience. For Martin Luther, salvation was by faith alone, not by good works or by social status: a very individualistic view. From then on, the pendulum swung back toward the individualistic pole. The Reformation created modern individualism, and individualism became the core feature of Western culture. In the twentieth century, the pendulum swung back to the collectivist ideals of communism and fascism. From a Chinese, 6,000-year perspective, neither of these lasted very long. Even China gave up persisting in the ideology of communism economically. Even so, this change could still be observed from the same perspective, swinging between individual and the group, since the modern Western democracy remains a collective way of life based on the individual. Every individual in the community is highly independent and self-aware. Community is no longer oppressive to the individual. The freedom of the individual is respected. The public affairs of the community are determined by the collective individual.

Western society has always swung back and forth between two extremes, the individual and the collective, because of the dichotomous mentality that originates from the linear mode of thinking. Based on this cultural background, Western interpersonal relations could thus only be individualistic and collectivistic unconsciously. When Hofstede (1980) created a value system to measure the global differences of culture, individualism was then an inevitable dimension. However, individualism implies collectivism. To Western scholars like Hofstede, their minds are still swinging between the two extremes. The measurement would then naturally be carried out in between, i.e., where the society locates between individual and the group. If the measured society situates at one extreme of individual, the society is comprehended as individualistic, or it will be collectivistic. Therefore, the key will be to find the boundary in between accepted by that society. The boundary means the range or limit which is meaningful to each extreme, and also the front line where the two extremes interact. When the boundary is very close to the end of the group, it means that the individual has plenty of space for thinking and behaving. It is an individualistic society, and so are interpersonal relationships. If the boundary is very close to the end of individual, representing very limited freedom for the individual, sometimes even intruding into personal private space, this society is called collectivistic as defined by Western individualism.

Obviously, whether from a perspective of individualism or collectivism, it is surely not appropriate to describe Chinese society and interpersonal relationships as well. The Chinese culture is a relationship-based ethical culture. Therefore, the ethics in Chinese culture are really the interpersonal ethics based on relations with family members. Because of the powerful influence of family, every Chinese person has to live among multiple relationships throughout their whole life. Except for oneself, everyone must learn to manage all sorts of different relations. The different relationships based on family ethics cannot be simply dichotomized into individual and the group, because in Chinese minds, there are many different kinds of group. Hence, the Chinese society is neither individualistic nor collectivistic (Yang, C.-F. 1993), and the same applies to interpersonal relations. The most fundamental reason for this mentality is thought to be from the holistic view from the ideographic mode of thinking, so as to emphasize and further differentiate those different relations with different persons.

The guiding philosophy of Chinese culture is Confucianism. The core concern of Confucianism is the establishment of supreme harmony in a society with humans at the center and without regard to any kind of god. In Confucian philosophy, no one is totally independent of others. Instead, everybody's existence is defined socially and interactively. Consequently the concept of "ren," emphasizing the relationships between people, was created and then became the core of Chinese culture. "Ren" refers to two persons and suggests the ultimate goal (or model) for human relations between any two persons. Specifically, "ren" is about interpersonal relationships. Everyone will have to live among different kinds of interpersonal relationships throughout one's life. "Ren" is then referred to as the basic belief of these relations. Confucianism further modified the five most important basic interpersonal relations in the traditional society to be:

- father and son;
- ruler and subject;
- husband and wife;
- elder and younger;
- friends.

Confucianism also offered five corresponding standard norms for each relation. They became the formal Five Cardinal Virtues, referenced above, of *affection* (between father and son), *righteousness* (between ruler and subject), *distinction* (between husband and wife), *order* (between elder and younger), and *sincerity* (between friends). Therefore, the core concern of Chinese culture, for thousands of years, is really about the interpersonal relationship.

The core of Confucius teaching is "ren" and it is at the heart of Chinese culture (Liang 1963). "Ren" also explains why Chinese culture is ethical. Although there were five basic interpersonal relations deduced from "ren", it is generally agreed by many Chinese scholars that there exist two cognitive dimensions to describe and categorize Chinese human relations (Chuan and Yang 1997). They are superior–subordinate and intimacy–estrangement. The dimension of superior–subordinate refers to any relationship having *position difference*, while the dimension of intimacy–estrangement refers to any relationship having *closeness difference.* No matter in which one of the five basic relations, both sides involved should resolve the interacting manner based on the superior–subordinate or intimacy–estrangement between each other. Actually, it was because of these two cognitive dimensions that ancient Chinese scholars could propose the corresponding values of the five basic relations in Confucianism, which became the Five Cardinal Virtues. It was also recognized as the reason why Chinese scholars did not propose any value corresponding to strangers since it was not possible to identify the position difference and closeness difference between the self and strangers; we simply did not know who the stranger was. This became a long-lasting characteristic of Chinese culture, that no values were affixed to strangers. The other reason for this phenomenon is that there is no stranger in a family.

The dimension of superior–subordinate is a concept much greater in scope than leader–follower. Jing et al. (2002) have described in great detail Chinese authoritarianism as "dragonality," which is a concept also much larger than power distance. "Dragonality" is useful for understanding Chinese authoritarianism wherever there is position difference.

In the dimension of intimacy–estrangement, the corresponding values were established based on closeness. Therefore, the "ren" proposed by Confucianism actually means close to those who are close. We should make ourselves close to the one close to us. Parents are very close to us, so we should behave as if they are very close to us and treat them as close as they are. Strangers are very far from us, so we should treat them as far as they are. That is to say, the manner we use to treat people should depend on their closeness to us. Different people with different closeness should be treated in different ways. We cannot treat different people all in the same way. It is very easy to understand from the perspective of family. We simply cannot treat our parents in the same way as we treat our brothers or sisters. Hence, the "ren" originated from Chinese family ethics is truly a system of values adhering to differentiated closeness.

There have been many theories and discussions about the relations between people in Chinese history. Fei Xiao-Tung (1947) was the first scholar to use Western scientific methods, field experiments, to investigate traditional Chinese human relations and came up with a new theory. He spent several years in Yunnan, a southwest province of China, investigating the traditional human relations among rural peoples. Finally, he proposed the concept of differentiated order, as shown in Figure 7.1. In his theory, Chinese human relations are modelled as a series of concentric circles, with the "self" at the center and people with different levels of intimacy, or degrees of closeness, at different distances from the center, like the concentric circular ripples caused by dropping a stone into water. People located in different rings are treated differently. The farther people are placed from the center, the less concern one has for them. In Chinese society, everybody is at the center of their own differentiated order and also at different positions in the concentric circles of other people.

By the differentiated order of Fei Xiao-Tung, the interpersonal relations in Chinese society all started from self, the center of all relations, using family as the prototype, pushing outward and forming different kinds of relations. Based on the primary relations in family, all social relations were just the generalization of different relations with family members. Therefore, the structure of Chinese society was then a network connected by countless families with different levels. Because of this, the fundamental differentiation would exist naturally between family members and non-family members. This was exactly the first differentiated order proposed by Fei: family members–non-family members, or simply family–outsider. Fei used this model to explain Chinese interpersonal relations and the corresponding interactions with great success.

In addition to this, Fei also mentioned that the boundary between family members and strangers was not fixed. It would move according to the situation confronted,

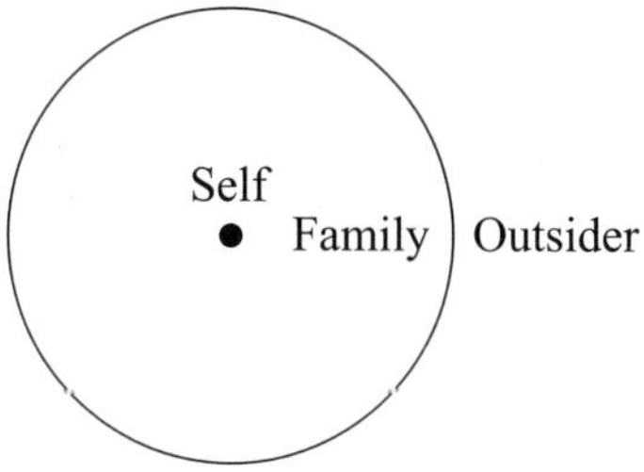

Figure 7.1 Fei's differentiated order: perceived interpersonal relations based on family from the Chinese ideographic mode of thinking

which means that the Chinese differentiated order was a flexible adaptive structure. This was also understood as situationalism (Hsu 1970). Americans accept that all humans are created equal and should be seen as individuals. For the Chinese, everyone is a member of a group in a different situation. Chinese culture admits no one is really independent from others. Everybody should rely on others, especially family members. From the very beginning, starting from the parental reliance in the family, more generalized, different kinds of mutual reliance were developed. Hence, it is not possible to create the concept such as human rights conferred by God, since everything is defined from interactions with others. Everybody should behave according to the consideration about different positions, relations, and closeness of others in different situations. In other words, the Chinese people were accustomed to suppressing their own feelings, preferences, and sentiments, and behave differently in different situations.

The concept of differentiated order has been widely accepted as the basic model to portray Chinese human relations and was further extended by several scholars. Ho and Chen (1991) agreed that the Chinese can differentiate human relations according to the situations they face, but they tend to use a relatively fixed conceptual frame rather than a dynamic one varying with the situation faced. Ho (1998) proposed the concept of "relationship dominance" to describe the importance of human relations as viewed by the Chinese. He also argued that the Chinese personality was heavily influenced by interpersonal relations. King (1992) acknowledged that "guanxi," favor and face are the key cultural concepts to comprehending Chinese society. He used the term guanxi network to describe Chinese human relations. Every Chinese would do their best to create their own network as a resource to ensure that their goals can be achieved given their social resources.

K.-S. Yang (1993) defined relationship orientation as the core operation style in Chinese human relations, and further named the three different types of relations as "family," "acquaintance," and "alien" or "stranger." Yang also discussed the interactions of these relations. Huang (1995, 1998) used the social exchange theory which proposed favor and face to explain Chinese social exchange behavior in the interactions of interpersonal relations. He further separated human relations into "sentimental," "implemental," and "mixed" types. People use different manners to deal with different types of relations. C.-F. Yang (2001a) divided Chinese human relations into "existing" and "interacting" components. "Existing" refers to the results of previous interactions, while "interacting" refers to the accumulated situation from ongoing manipulation. Yang also argued that the Chinese people categorized interpersonal relations according to these two scales and interact with different rules.

Gau and Chen (1989) performed research using local companies to study social relations. They proposed the trust concept and described trust as the core of their system for classifying Chinese relations instead of using kinship or something else. Yang (2000) argued that Chinese human relationships not only emphasized the different roles to play but also stressed the position difference in the corresponding roles. Yang further discussed that interpersonal relations do not mean a fixed group of people. It is only a range, a flexible and permeable range. Family members could become strangers, and strangers could also become family members. There existed a core category proposed by Fei as the most popular prototype of interpersonal relations, which is family–outsider.

From these discussions of the literature, the differentiated order by Fei was accepted to be the most appropriate concept of Chinese guanxi. The only classification mentioned by Fei is family–outsider, with a variable boundary. Although other more complex classifications exist, they lack supporting, scientific evidence. However, there is a survey which supplies some numerical results. Chuan and Yang (1997) performed a role-playing study by using superior–subordinate and intimacy–estrangement dimensions as part of a survey with more than 100 respondents. Their study listed 28 different human relations. They found that there are no distinct classifications, and that guanxi is continuous. The relations they used for the study are teacher, student, neighbor, college friend and many others. However, in the minds of Chinese people, these are all nominal variables. A friend could be treated as a member of the family sometimes, but as an outsider in other situations. Consequently, the results tend to be continuous if relations such as friend are used to measure the degree of intimacy–estrangement. Also, from the trust concept proposed by Gau and Chen (1989), trust was found to be the core concept for classifying Chinese relations instead of kinship or something nominal. Thereafter, guanxi cannot be classified or understood by formal names without the connotation of trust. However, trust is an appropriate concept to describe the type of treatment, but it is not a description of the structure of Chinese human relations (Cheng 1995).

There is no doubt that human relations are very important in Chinese society. Also, everybody accepts differentiated order with a dynamic boundary to be the most appropriate model to portray Chinese human relations. It is also true that there is no scientific evidence to reveal the structure of the classification of Chinese interpersonal relations. According to Fei himself, the concept of differentiated order really came from the basic meaningful unit of Chinese society: family. Therefore, family is still the prototype to understand the operation of Chinese human relationships, although the boundary of family varies according to different situations.

Chapter 8
Guanxi Gradient

The most important influence on interpersonal relations in flight safety is the interactions that occur among crewmembers. In the global aviation community, crew resource management (CRM) has become synonymous with flight safety. No matter how CRM is conceived globally, synergy is at the core of CRM. How the crewmembers work together affects the performance of the crew as well as the flight safety. In a cockpit, there is only one captain and one first officer. The captain is expected to take the lead and give orders while the first officer is expected to follow and take the orders. Therefore, the main concern in a cockpit is then the power operation between the crew. Because of the low power distance inherited in the basic design of CRM, as CRM prevails over the world, the primary training of CRM usually falls into the realm of lowering the power distance.

The cultural dimensions used by Hofstede (1980) to measure the differences among nations are power distance, individualism, uncertainty avoidance, and masculinity. Power distance and individualism were identified to be the most important ones. Helmreich (Helmreich and Merritt, 1998) found that the strongest differences between pilots of different countries were observed to be the command roles and interactions, attitudes toward orders, rules, and routines, and preference for and reliance on automation. A correlation study was also executed. The correlations between the three items and power distance were found to be 0.79, 0.67, and 0.74, while the correlations with individualism were 0.86, 0.88, and 0.74. From the data, the strongest differences between pilots of different countries were more correlated with individualism. In other words, if we are going to choose one Western value or belief to explain statistically the difference among pilots from different nations, individualism will have the highest explanatory power. In this sense, interpersonal relation is a cultural concept more fundamental than power or authority. At least, how international pilots perceive authoritarianism will have less impact on flight safety compared with interpersonal relations.

From a perspective of interpersonal relations, in Chinese minds, everybody possesses a central position surrounded by everyone else. According to closeness, or the degree of intimacy, every other one is placed a certain distance from the center. The whole Chinese society is then a huge network composed of infinite basic concentric structures. It is called a differentiated order. Differentiated order is a fundamental description of Chinese interpersonal relationships. Based on family ethics as the prototype, all other people are located at different psychological distances according to the degree of intimacy. Just like dropping a stone into the water, the ripples move from the center to the furthest distances. It also decides the basic pattern about how the Chinese exchange their social resources.

The person located closer to the center will be treated more positively. As the psychological distance becomes larger, the attitude decays. Hence, anyone placed further from the center will then be treated indifferently.

When the Chinese encounter different situations, family is always the basic framework used to divide close from remote, and intimate from estranged, in order to manipulate the social resources effectively, which means with the greatest possible expectation. The people close enough to us are usually treated as family members. In different situations a network very similar to the family clan is formed, which then supplies care and help for the members, including the exchange of sentiment and various benefits. People outside the rim of the family are basically treated as aliens, or at least as strangers. The exchange of sentiment and benefit is downgraded to be negligible, even negative. The criterion of judgment to divide family from outsider is based on situation. Under different situations, the envelope of family members will be varied.

The situations we have to face are simply too diverse. If different criteria are to be used to categorize so many situations, there will be too many of them. Human minds prevent us from using overly complicated categorizations. Generally people simply cannot memorize too many divisions. Hence, there must be a sufficiently simple and efficient classification to differentiate interpersonal relations so as to effectively manipulate social resources. This is obviously more complicated than individualism and collectivism.

The data from the large-scale survey about CRM in Taiwan discussed in Chapter 6 (Lu et al. 2000) is used again as the source of scientific data about Chinese differentiated order. The goal of this study was to explore the face of Chinese interpersonal relationships derived from family, especially differentiated order, in the cockpit. Those questions with answers significantly different between the Taiwanese and international pilots were grouped again. Since the study was focused on interpersonal relationships, the questions apparently about the perception of intimacy and estrangement were collected, which means that of the 88 questions the pilots answered, only the questions directly related to human relation are chosen for analysis here. Most of the questions used a five-point Likert scale, representing strongly agree, agree, neutral, disagree, strongly disagree, or very frequently, frequently, sometimes, seldom, very seldom. From 5 points to 1 point were assigned to each answer. Means and standard deviations were obtained. The student's t-tests were also used to check for significant differences in the answers between the local and foreign captains and also between different categories. The significance level chosen here was $\alpha = 0.05$. In total, there were nine questions whose results for local and foreign captains differed significantly. All of these questions are listed in Table 8.1 according to significance level.

The first question related to human relations with the highest significant difference between the local and foreign captains is question 34, "The pilots I fly with in my company are good communicators." The average for the Taiwanese pilots is 3.83 and the corresponding average for the international pilots is 3.34, with a t-value of 8.00. The difference between them is quite significant.

Table 8.1 Survey questions relating to interpersonal relations

N	Questions	Local Captain		Foreign Captain		t-test	
		Mean	Standard deviation	Mean	Standard deviation	t-value	p
Q34	The pilots I fly with in my company are good communicators.	3.83	0.59	3.34	0.82	8.00	***
Q35	The pilots I fly with in my company maintain good situation awareness.	3.87	0.55	3.52	0.95	5.52	***
Q38	The pilots I fly with in my company manage the workload effectively.	3.96	0.57	3.71	0.82	4.14	***
Q40	The pilots I fly with make appropriate decisions in a timely manner.	3.86	0.58	3.69	0.77	2.87	**
Q36	The pilots I fly with in my company are good team players.	3.84	0.61	3.69	0.80	2.53	*
Q65	Whenever there is conflict, people in my company are more concerned with what is right instead of who is right.	3.45	0.94	3.18	1.11	2.98	**
Q55	The people I work with faithfully follow the company's standard operating procedures.	4.11	0.62	3.97	0.86	2.17	*
Q18	I will make a special effort to understand crewmembers from other nationalities.	3.52	0.86	4.33	0.74	-11.11	***
Q19	I like to work with pilots from different nationalities.	3.08	1.04	3.84	0.82	-8.87	***

Notes: * $p<0.05$; ** $p<0.01$; *** $p<0.001$ (5 = "strongly agree," 4 = "agree," … 1 = "strongly disagree").

The local captains agreed that the pilots they fly with in the company can communicate very well with them. The percentage of the foreign pilots that agreed was not as high as the local pilots. They were more close to neutral on this question. There is one thing we have to remember. No matter for local or international pilots, they were all employees of Taiwanese airlines. The pilots they fly with were actually the same group of pilots working in Taiwan. The same group of pilots could not possibly have two different levels of communication performance. Hence, the difference could only come from the different perceptions between local and foreign pilots. Facing the same group of pilots, Taiwanese captains accepted they are good communicators and the foreign captains perceived they were just neutral, and the difference was significant by the t-test. It indicated that the Taiwanese pilots inclined to give more positive comments to their colleagues than the international pilots.

The question with the next highest t-value is question 35. It was still about the pilots flying with them: "The pilots I fly with in my company maintain good situational awareness." This question was almost the same as question 34 except that it asked about situational awareness instead of communication. The average score of local captains was 3.87 and international pilots 3.52. The t-value was 5.52. The difference was still very significant statistically. The meaning of results of this question is almost the same as question 34. To the same group of pilots, the local captains provided better judgment about their situational awareness. Again, the Taiwanese pilots valued more about their colleagues than the international pilots.

It was not only the above two questions that showed these relatively distinctive results; there were three more questions that also exhibited the same phenomenon.

38. The pilots I fly with in my company manage the workload effectively.
40. The pilots I fly with make appropriate decisions in a timely manner.
36. The pilots I fly with in my company are good team players.

The averages of question 38 for the local and international captains were 3.96 and 3.71, with a t-value of 4.14. Again, the Taiwanese pilots gave much better comments about their colleagues. The means of question 40 for both sets of pilots were 3.86 and 3.69, with a t-value of 2.87, and the means of question 36 for both sets of the pilots were 3.84 and 3.69, with a t-value of 2.53. From the above results, it was clearly found that the local pilots consistently supplied wonderful opinions about their workmates, and agreed that they are good communicators, they maintain good situational awareness, they manage the workload effectively, they make appropriate decisions, and they are good team players. Almost all the possible good descriptions about the pilot were attached to the pilots used to fly with them.

Among those questions with significant difference between the local and foreign pilots, there were two other slightly different questions. The first one is question 65, which asked about people in the company instead of the pilots they fly with.

65. Whenever there is conflict, people in my company are more concerned with what is right instead of who is right.

Actually, this question was asking whether to allow personal judgment to intrude, or not. The mean of the local captains was 3.45 while the result for foreign pilots was 3.18, with a t-value of 2.98. From the above results, it was found that the Taiwanese captains still supplied relatively good comments about people in their company, but the difference was not as significant. In reality, Chinese people get personal much more easily because almost every Chinese person is heavily surrounded by complicated interpersonal relationships. It is then quite hard for the Chinese to make considerations without referring to somebody. That is to say, the difference should be much smaller than the data revealed. And, the result from the international pilots, 3.18, is thought to be much more objective and closer to the reality. The percentage of people in the company more concerned with what is right instead of who is right was only a little bit above the average.

There was another question asking about the people in the company.

55. The people I work with faithfully follow the company's standard operating procedures.

In the previous questions, it was clearly mentioned about the "pilots I fly with." In this question, only "the people I work with" were mentioned. Hence, it could be accepted that "the people I work with" were not the pilots I fly with. Further, the psychological distance for "the people I work with" was obviously larger than the pilots I fly with. The average of the local pilots was 4.11, while the average of foreign captains was 3.97, with a t-value of 2.17. The local pilots still provided better remarks about the people they work with than the foreign pilots as expected. However, it was right because the Taiwanese pilots not adhering to the standard operating procedures cause about 10% more accidents. Again, it is believed that the results from the international pilots were more representative to the fact in Taiwan. From the above two questions, it was again found that the Taiwanese pilots still provided much better comments even about the people in the company instead of just about the pilots who flew with them.

The first five questions (34, 35, 38, 40, 36) were all related to the pilots flying together. The questions concerned whether the pilots have good communication skills, maintain good situational awareness, handle workloads efficiently, make appropriate decisions, and are good team-workers. The scores from local captains are all significantly higher than those from international captains. Further, the standard deviation of each question is consistently lower. These results strongly suggest that the way Chinese captains treat the pilots flying together is very positive from that of international captains. As for the next group of questions, 65 and 55, they are about colleagues in the company, whether they follow SOPs and whether they care more about what is right or who is right. Again, the scores from the local captains are higher, the standard deviations are also consistently lower, and the differences are significant, which means that the local pilots still provide quite positive comments about them.

However, the t-values of questions 65 and 55 are lower than the t-values of the first group, i.e., questions 34, 35, 38, 40, and 36. This has an important implication: compared with the international reference, the local captains treat pilots flying together differently from the colleagues in the company. To prove this, one more t-test was executed. All the data from questions 34, 35, 38, 40, and 36 were collected together as the first group, with data from questions 65 and 55 as the second group. The comparison was made between the difference from local and international captains of the first group and the difference of the second group by using the two-sample t-test. The t-value was found to be 1.693 greater than t_{crit}. Again, the difference between the first and second group was significant. Obviously, in the minds of Taiwanese captains, pilots flying together are psychologically closer than their colleagues in the company with reference to the international average.

In Table 8.1, among the questions with significant difference, there were two more questions asking about pilots from different nationalities. They were listed as followed. The first one is question 18. It stated:

> 18. I will make a special effort to understand crewmembers from other nationalities.

The mean for the Taiwanese captains was 3.52 and the corresponding number for the international pilots was 4.33. This time it was the mean from foreign pilots that was larger than the local pilots. From the number, it could be said that the local pilots took a roughly neutral position to make a special effort to understand crewmembers from other nationalities. The international captains were more than agreeable to make such a special effort. The difference was strikingly significant, with a t-value of 11.11. Hence, compared with the international captains, it is acceptable to say that the Taiwanese pilots were generally not willing to make a special effort to understand crewmembers from other nationalities. To the foreigners, it would seem to be quite hard to understand why it occurred. However, to the Chinese, it is so natural that there is no need to even say anything about it. What need is there to understand crewmembers from other nationalities? There is no reason for doing that. Superficially, it seems to be explained by Chinese utilitarianism. Actually, it is not. The fundamental reason for this behavior is that the crewmembers from other nationalities were out of the scope of Chinese ethics derived from family. The Chinese simply did not know how to treat these "people" since there is no category for them in Chinese minds. For Western culture, all these strangers were just neighbors, and you should love your neighbors. If you could love your neighbors, of course you will make a special effort to understand them.

The second question asking about pilots from different nationalities was question 19.

> 19. I like to work with pilots from different nationalities.

The average of the question for the local pilots was 3.08, while it was 3.84 for the foreign pilots, with a t-value of 8.87. The t-values of questions 18 and 19 were the largest among all the questions studied in this work. If the Taiwanese pilots did not like to understand the foreign pilots, of course they would not like to work with them. It was not surprising that the Taiwanese captains again took a neutral position in this question, and the difference was also very significant – as the previous one. If I do not know how to categorize these people, how is it possible that I like to work with them? Together with the previous question, it was obvious that the pilots from different nationalities could hardly find a position in the Chinese minds. They were located just too far away from the Chinese "self," so that the images of them became blurred.

Putting all these nine questions together, it was found that they could be categorized into three groups, since statistically the results were different significantly. The three groups are: pilots I fly with, colleagues in the company, and pilots from other nationalities. According to the Chinese differentiated order, the pilots I fly with are people working with me in a very narrow space, accomplishing many flight missions, even cooperating together to overcome certain difficulties. The psychological distance would naturally be very close. At least, those pilots must be very familiar to us. As for the colleagues in the company, most of them would be unfamiliar to me. The psychological distance would then be much further relatively. However, since we are all working in the same company, we surely have certain connections, and of course we share something in common. If the company is just like a big family, we are all family members, no doubt about that. Talking about the pilots from other nationalities, I do not know them. Further, after they finish their contracts, they will go back to their countries and I will never have any chance to meet them again. Where could I place them in my mind psychologically? They would be just too far away such that the family concept simply could not be applied to them. Under such circumstance, why should I make a special effort to understand them, and why would I like to work with them? What is the meaning of doing that? According to the Chinese ethics from family, they were just like strange visitors to our family. All we have to do is to treat them politely according to the etiquette without any affection.

It has been shown that the ways that local captains treat people in different psychological distances are quite different. From the results, pilots flying together locate at the range different from that of colleagues in the company. The pilots with different nationalities are situated at a range even further away. Compared with the international captains, this represented an expectation for the local captains used to categorize human relations and manage it as resources. Obviously, from a more fundamental view, it delineates a mental model that existed in the minds of Taiwanese captains.

The differentiated order is an appropriate description of Chinese human relationships. It represents how Chinese categorize the degree of intimacy between different people. Although there are several models describing the basic structure of the differentiated order, a lack of numerical evidence makes them difficult to justify. The original goal of the CRM survey in Taiwan was not to study the differentiated order, although the results do reveal a clear structure. In Table 8.1 we can see that local captains clearly place pilots flying together, colleagues in the company, and foreign pilots at different locations. Also, local captains treat each of these groups differently. Hence, there exists a clear structure of how local captains categorize and treat people. When referring to pilots flying together, local captains give much more positive descriptions, e.g., have good communication skills, maintain good situational awareness, etc. Thereafter, the pilots flying together can be grouped as a category. To the local captains, they represent close colleagues with many chances to work together, to cooperate. "Acquaintance" is used to represent this category in the present study.

As for the next category, colleagues in the company, local captains treat them differently than pilots who fly together. They still give positive comments, e.g., they follow SOPs, but not as high evaluations as those given to pilots who fly together. For colleagues in the company, although they are people unknown to the captains, they may still be perceived as having the same interests, beliefs, goals, etc. "Fellow" is thought by the authors to be the appropriate English name for this category. From the results in Table 8.1, it is clear that local captains are not so willing to work with foreign pilots. Although they may have the opportunity to work together, and they are possibly familiar to them, they still emotionally locate them at a far more remote distance than colleagues within the company. It would appear that local captains do not really care about the foreign pilots. In accordance with the t-test results, one more category is needed to represent this class. "Alien" is used in the present study.

In the Chinese differentiated order, people located closest to oneself should be related by blood or marriage, in other words, a biological family. Even acquaintances cannot be compared with these people, since they are inherited. To the Chinese, family implies obligations rather than resources to manage. Actually, in Chinese, "family" bears a wider meaning. Hence, "kin" is a more appropriate name to represent the category of people closer than acquaintance.

Based on these previous studies, Jing (2004) proposed a four-level differentiated order to describe the basic mental structure of Chinese human relations: kin–acquaintance–fellow–alien. With self at the center, "kin" refers to the people located closest to the self and related by blood or marriage. "Acquaintance" represents those people quite familiar to us, up to a status that we can have sentimental and beneficial exchange with them, like friends, neighbors, and colleagues. People not very familiar or even not known to each other, but who still have certain things in common, for example working in the same company, graduating from the same university or joining the same party, belong to the category of "fellow." As for "alien," it represents the people we do not want to work with or even to understand – those who are psychologically remote. Thus kin–acquaintance–fellow–alien is a four-level model of the Chinese differentiated order. It was derived from the concrete scientific evidence of the present study. The survey of CRM about the Taiwanese pilots and management level clearly revealed the structure of interpersonal relations in Chinese minds. It is the natural extension of the family–outsider distinction of Fei (1947). Fei's differentiated order was from the perception about the family from Chinese dragon culture, and as the basic prototype of all organizational operations in Chinese society.

Kin–acquaintance–fellow–alien is a four-level model of the Chinese differentiated order. Although each level represents a specific category of intimacy or trust between people, it is not a fixed structure. For example, an old friend belonging to "acquaintance" can be regrouped into the "kin" level through marriage.

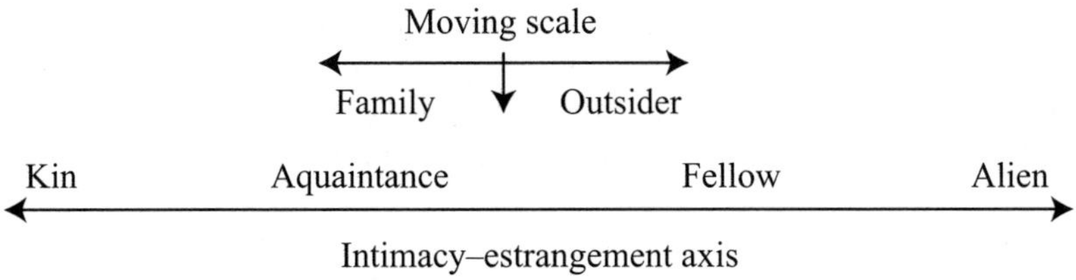

Figure 8.1 The four-level differentiated order together with the moving scale of family–outsider to portray the Chinese concept of interpersonal relations

A relative can also be relocated as alien under some circumstances. The four-level differentiated order only provides a basic understanding of the complicated Chinese concept of *guanxi*. Further, it is still not enough when facing certain social situations by using only such a static structure. To help manipulate the resources of human relationships, there should be a moving scale for Chinese to optimize management. Since family is the fundamental concept of Chinese human relations, it is used almost everywhere to classify people. It is also the original classification used by Fei (1947), family–outsider. As it is also mentioned, the boundary between family and outsider is movable and permeable. In the present study, it is used as a moving scale in addition to the proposed basic four-level structure. Combining the kin–acquaintance–fellow–alien four-level basic structure and the family–outsider moving scale as shown in Figure 8.1, it was believed to be a fairly complete picture for the Chinese differentiated order.

To have a clear picture about the safety-related implications of the differentiated order, the case discussed in Chapter 2 can be used here to reveal the possibility of deterioration of cockpit operation designed by Western individualistic culture. In January 1995 on the Chinese lunar New Year's Eve, the last local flight back home crashed on a hill close to Taipei Airport. Two flight crew and two cabin crew were killed. According to the accident report, it was found clearly that a stewardess was sitting in the cockpit. The flight started at 19:13 and ended at 19:43. According to the voice recorder, the atmosphere in the cockpit was really relaxed and pleasant. During this period of flight, the stewardess spoke 28 times, mostly asking questions, and the captain responded 21 times, while the first officer responded 3 times. The first officer was the flying pilot and the captain was monitoring. At 19:41, the captain spoke to Taipei approach, "airport in sight," and Taipei approach responded with "cleared for visual approach." That means the captain could already see the airport. At 19:42, the stewardess asked the captain about the appropriate time to lower the landing gear. At the same time, the air traffic controller (ATC) also called into the cockpit and asked the captain about the aircraft speed twice. The captain answered the stewardess first and then responded to the ATC. At 19:43, after the first officer called "capture" for an unknown reason, the heading changed together with the roll angle. The captain said "No!" twice.

Four seconds later, the captain said: "Sir, don't deviate." It is quite clear that the captain knew it was wrong but did not stop the first officer from deviating away from the approach. Only 13 seconds after the captain spoke, the recording ended. Before the end, it was recorded that the aircraft was just commencing the process of pull-up. Unfortunately, it was too late.

From this accident, it is quite clear how Chinese differentiated order influenced the crew performance in the cockpit. From this example, the first thing in the cockpit to be distorted is communication. If someone viewed as alien is in the cockpit, an ordinary Chinese pilot tends not to intimately interact with him. The consequence will be a quiet cockpit. On the other hand, if a very close acquaintance works in the same cockpit, the atmosphere likely becomes easy and relaxed. Too much talking not related to the flight operation was found quite often. Once in a while, even the stewardess can be found in the cockpit, as in the above-mentioned case. Sometimes even the normal communication with ATC was interrupted. All of these could be understood by differentiated order since the Chinese treat people at different psychological distances differently. The stewardess might very likely be treated as acquaintance by the crew and be able to share so many things with the crew. On the contrary, the air traffic controller was obviously placed at a further distance by the crew compared with the stewardess.

The second thing to be influenced is monitoring actions. In the cockpit, when one pilot is flying, the other one should be monitoring. It can be found from the above accident that the captain talked too much to the stewardess. It surely will interrupt the monitoring duty assigned to him. When confronted with a very close acquaintance, it will generally be very hard to maintain a strict monitoring attitude for an ordinary Chinese. If the relation of superior and subordinate existed, the situation would deteriorate even more. Hence, it is reasonable to question whether the Western-oriented manner in the cockpit based on the relatively individualistic human relation is vulnerable in certain cases to the differentiated order.

One may ask why the pilots do not just simply ask the stewardess to leave the cockpit. In the Chinese society, human relationship is a valuable resource as well as an unavoidable social pressure. Once the pressure has been built up, i.e., the close relationship is established, it is generally not easy for most Chinese people to resist. The main reason is that if the captain asked the stewardess to leave, their good relationship would very likely deteriorate and even be downgraded. From the resource point of view, it represented a loss of valuable resource which most Chinese would not do. Hence, if the cockpit manner used today is not to be modified in the near future, it is important to bear in mind that the Chinese differentiated order must be faced with care. The key to efficiently manage the Chinese differentiated order in the cockpit is to relieve the pressure in the minds of pilots from close human relations. There are at least two ways to achieve part of this goal. The first one is not to put two pilots with very close relations into the same cockpit, although it will be rather difficult to do so. The second will be to make the rules clear. It will give the pilots a likely excuse to help them resist the pressure.

This psychological structure was deduced from the unprecedented survey of CRM in Taiwan. If we try to use other categories, such as friends, colleagues, classmates, etc., the explanatory power is limited since the differentiated order originated from Fei was not a category by the names of human relations. Instead, it is an inner structure deep in Chinese culture about intimacy and estrangement, sometimes even including trust. Such a psychological structure surely could not be understood by the simple dichotomy of individualism–collectivism.

The Western concept of human relations is basically dichotomous because of linear thinking. In their minds, Chinese human relationships are surely not individualistic, so they must be collectivistic. Of course this is not correct. The Chinese human relationship can never be understood by any kind of linear concept. Because of the ideographic mode of thinking, the human relationship is perceived by the Chinese as a picture with different persons at different positions, like a solar system, not a line with only two ends. According to the Chinese differentiated order, the Chinese treat different people located at different positions differently. Moreover, the further away the person locates, the worse the attitude becomes. In other words, the attitude decays as the psychological distance increases. Hence, if the tendency of this decay could be used as a cultural variable, it would be expected to be more useful than the simple Western dichotomous individualism and collectivism as proposed by Hofstede (1980).

Before a new cultural variable could be defined, the complicated Chinese concept of "guanxi" must be explained clearly, since the differentiated order is only a model to depict guanxi. Guanxi is usually translated as "relation" or "connection," but it is much more than that. It is an attitude strictly concerned with facing people with different degrees of intimacy. This means that guanxi can be good or bad. The Chinese people treat guanxi as a form of social capital, or manageable resource for human relations. Since this capital is always limited, it will be better invested in persons with an expected high interest rate, usually family members or old friends, etc., and is gradually reduced with less intimate people. It is always intermingled with authoritarianism because the person with great power is the one we most like to establish good guanxi with, or this happens to be the one with whom we already have very good guanxi. In this case the expected interest will be extremely high. However, the guanxi is a resource as well as a pressure. When a good guanxi has been established, the pressure is also accumulated as well. The pressure comes from being afraid of loss of this valuable resource. Sometimes, when a request comes from a person with very good guanxi, e.g., spouse or parents, the pressure is almost impossible to resist. Frequently, when a guanxi network operation violates bureaucratic norms, it can lead to corruption, such as nepotism and cronyism.

Talking about a new cultural variable regarding attitude toward intimacy and estrangement as a supplement to authoritarianism, it should be a variable about guanxi. Combining the CRM survey data obtained in Taiwan and the four-level differentiated order model proposed by Jing (2004) – kin–acquaintance–fellow–alien – the "guanxi gradient" is defined according to the different attitudes used to treat different categories of people. It is hoped that the guanxi gradient can be used as another dimension in addition to authoritarianism, and can supply more information about global interpersonal relations than individualism to promote mutual understanding in the global flight safety community.

From the previous results (Jing 2004), the four-level differentiated order provides the possibility for quantification of the Chinese concept of guanxi. The data from the large-scale survey – the Human Factors and Safety Culture Survey (Lu et al. 2000) was used again. Of the questions the pilots answered, only nine questions directly related to human relations and thus these were the ones used for further analysis. On these items there were significant differences between the results from local and international pilots. These questions are listed in Table 8.1, ranked according to significance level. The first five questions (34, 35, 38, 40, 36) all related to pilots flying together and are grouped under the title of "acquaintance" according to the definition from guanxi. As for the next questions, 65 and 55, they are about colleagues in the company, who are represented by the category of "fellow" in the four-level differentiated order. When asked about foreign pilots, the results given in questions 18 and 19 clearly show that the Taiwanese captains seemed more reluctant to work with or even to understand such colleagues when compared with international captains. Obviously, they are grouped as being "alien" because the foreign pilots are psychologically pushed out of the limit of family by the Taiwanese captains. With these different groups of people representing different levels of intimacy, the horizontal axis illustrating intimacy–estrangement can be created. On this basis the study proceeded further.

In order to compare different attitudes from different cultural backgrounds, the international captains were further grouped according to region. There were 40 from North America, 21 from Latin America, 28 from Asia (excluding Taiwan), 32 from Europe and 14 from Australia. Since the goal of the present study was to explore human relations in the cockpit using the concept of intimacy–estrangement, a basis of comparison was needed because there was no universal standard measure for human intimacy. From the results, the data from Australian captains seemed to be generally the lowest compared with the other regions. Also, it is agreed that Australian people are very individualistic (Braithwaite and Caves 1997; Hofstede 1980). The data from Australia were then used as the reference point for comparison. The averages from different regions were all normalized with respect to the Australian data. That is to say, the averages from different regions were subtracted from that of Australia to obtain the differences and then divided once again by the same figure. The corresponding results, called normalized differences, are shown in Table 8.2 and are used for further calculations.

The different levels of intimacy can be established from the four-level differentiated order model, which can then be used as the horizontal axis. However, as the definition of "acquaintance," "fellow," and "alien" are relatively arbitrary, it could be argued that where they are situated on the horizontal axis could be designed to convey the appropriate meaning from the data. Thus, the first question in this study is how far away the concept "alien" is psychologically from the pilots.

Table 8.2 Normalized differences of the attitude toward facing different levels of intimacy for pilots from different regions

		Taiwan	North America	Latin America	Asia	Europe	Australia (baseline)
□	Sample	363–8	34–40	20–21	28	32	14
Q34	Average	3.83	3.2	3.45	3.57	3.53	2.93
□	Normalized difference	0.31	0.09	0.18	0.22	0.21	□
Q35	Average	3.87	3.35	3.95	3.75	3.59	3.00
□	Normalized difference	0.29	0.12	0.32	0.25	0.20	□
Q36	Average	3.84	3.53	3.71	3.82	3.78	3.29
□	Normalized difference	0.17	0.07	0.13	0.16	0.15	□
Q38	Average	3.96	3.73	4.09	3.71	3.75	3.43
□	Normalized difference	0.15	0.09	0.19	0.08	0.09	□
Q40	Average	3.86	3.38	4	3.64	3.81	3.36
□	Normalized difference	0.15	0.01	0.19	0.09	0.14	□
Q55	Average	4.11	3.85	3.90	4.18	3.84	3.79
□	Normalized difference	0.08	0.02	0.03	0.10	0.02	□
Q65	Average	3.45	2.70	3.52	3.82	3.31	2.86
□	Normalized difference	0.21	-0.06	0.23	0.34	0.16	□
Q18	Average	3.52	4.56	4.52	4.07	4.59	4.36
□	Normalized difference	-0.19	0.05	0.04	-0.07	0.06	□
Q19	Average	3.08	4.10	3.67	4.00	4.25	3.71
□	Normalized difference	-0.17	0.10	-0.01	0.08	0.14	□

To study this, on the horizontal axis representing intimacy–estrangement, self is assumed to be located at the origin. Kin, acquaintance, and fellow are located at points 1, 2, and 3 respectively, which is somewhat arbitrary. However, by doing this, it is then possible to show how distant the "alien" is from the concept of "fellow." Since there are no data for kin, the results showing the psychological distances from alien can only be understood as being normalized by the distance between acquaintance and fellow. We thus have data showing on the horizontal (acquaintance, fellow) and on the vertical (normalized differences of attitude) axes. The horizontal coordinates of alien were undetermined and were derived by calculation. For simplicity, only a linear fit was used to find the most appropriate

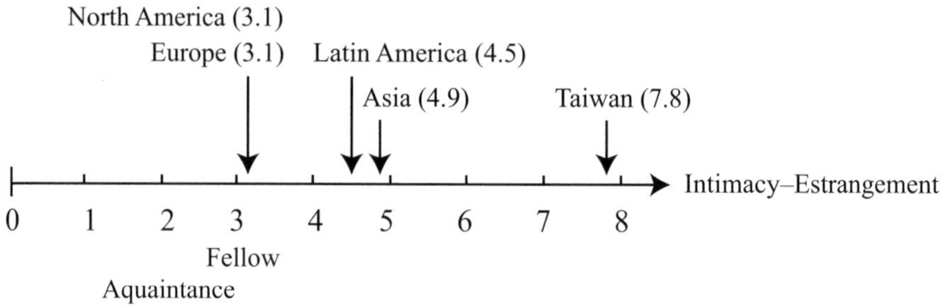

Figure 8.2 Locations of the concept of "alien" for pilots from different regions

position for the alien, i.e., the position of alien with the smallest mean square error when fitted using a straight line. The results are given in Figure 8.2. In this figure, it is quite clear that the position for "alien" represents how remote it is from the "self" as compared with the positions of "acquaintance" and "fellow." This figure reveals the striking differences between the pilots from different regions of their perceptions about aliens.

In Figure 8.2, the striking differences in the attitude toward "aliens" can easily be seen. Based on the distance of "acquaintance" and "fellow," the psychological positions of "aliens" are quite close to those for "fellow" for European and North American pilots. According to these results, it is thus reasonable to say that for Western pilots, there is almost no difference between the perception of "fellow" and "strangers," a natural outcome of individualism since they are all "others" besides the "self." Or, they are all neighbors. However, it is not the case for the other regions. For the local Taiwanese pilots, raised on the ethical values of Confucianism with an emphasis on the differentiation of interpersonal relations, it is very natural that the position of an "alien" is psychologically quite distant from that of "acquaintance" and "fellow." For the perceived "alien," in the long history of China, there exist countless examples showing the obvious indifferent attitudes toward people such as the enemy, rivals, traitors, and sometimes even foreigners or strangers. Although all cultures have negative views of these, a famous saying manifests the salient Chinese view: "the intention of the one not among us must be different." As for the pilots from Latin America and Asia excluding Taiwan, the positions of the perceived "alien" were located roughly in between these two extreme positions.

A further analysis was performed to explore the deep meaning of these data. In this case, the positions of the acquaintance, fellow and alien are fixed at positions 2, 3 and 4 on the horizontal axis. The vertical axis still represents the normalized differences. Again, a linear best fit is used. The results when these data are fitted with straight lines are shown in Figure 8.3. The corresponding best-fit equations are given below:

Taiwan	g(x) = -0.1835x + 0.6002
Latin America	g(x) = -0.0921x + 0.3902
Asia	g(x) = -0.0759x + 0.3138
Europe	g(x) = -0.0333x + 0.2173
North America	g(x) = -0.0035x + 0.0773

Different regions possess obviously different slopes. The slope here represents the rate of change of attitude with different degrees of intimacy. It also implies how the pilots from all over the world differ in their perception about human relationships. It is believed that this is basic human nature to have a type of gradient in attitude when dealing with people located at different psychological distances. In addition, this gradient must be negative, i.e., there is a positive attitude to closer people, such as family members, and a less positive, even negative attitude, to less close people, such as strangers or the heathen. This gradient can also tell how people from different cultural backgrounds manage their interpersonal relations. Moreover, since the guanxi for the global pilots is always decayed according to the different degrees of intimacy, the gradient type of guanxi appears invariably. Consequently, it is referred to as the "guanxi gradient" (Jing and Chang 2006). It is suggested that this is a useful concept for the cross-cultural study of global interpersonal relations.

From Figure 8.3, the "guanxi gradient" is defined as the gradient of the declining attitude toward different levels of intimacy concerning interpersonal relations. It is quite obvious that, for North American pilots, the gradient is almost flat given the reference from Australia. It is hard to tell the difference between them.

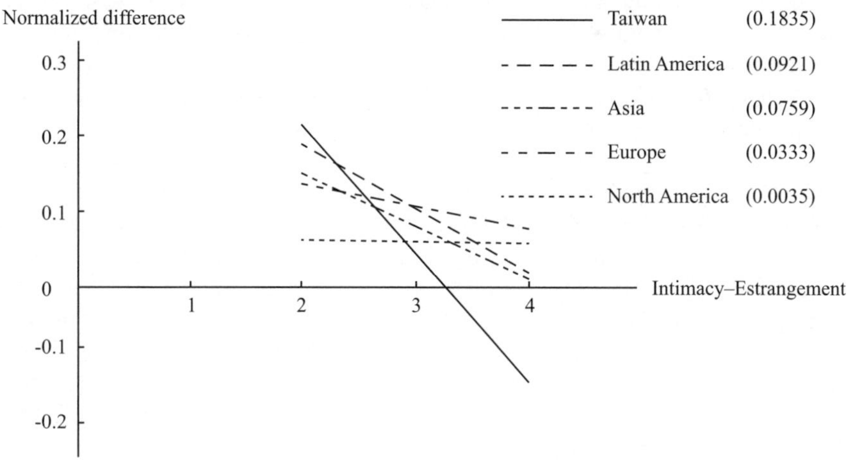

Figure 8.3 The international "guanxi gradients," the decay of the attitude according to the changing degree of intimacy, for pilots from different regions

Basically, Western culture believes all people are created equal. In their minds, everyone establishes their own relationship with God. Except that, all other people are just neighbors. It is not necessary to further differentiate the neighbors into different neighbors. Further, this can be considered as a strong indication of individualism. When compared with Hofstede's individualism index (1980) it is found to be in fairly close agreement, with the results showing that North America and Australia possess the highest scores amongst the other nations for individualism. As for the European countries, the guanxi gradient, slightly larger than 0.03, is higher than North America. The attitude toward aliens with respect to acquaintances of them became different. It was also found in Hofstede's individualism index that European countries possessed the second highest on the list just lower than North America and Australia. All of these countries were called high individualism countries by Hofstede.

On the other hand, by looking at the guanxi gradients for those countries with lower index values, some inconsistencies can be observed. In Hofstede's list, the countries from Latin America and Asia were mixed together and possessed lower positions on the list. They were what Hofstede called the lower individualism countries. Their guanxi gradients were found to be around 0.08 for Asian countries, with Taiwan excluded, and 0.09 for Latin American countries, which were still roughly consistent with Hofstede's index. The most striking difference was found in Taiwan. Taiwanese pilots possessed the steepest gradient of guanxi in the whole world. It is close to 0.2. Actually, it should be recognized as Taiwanese, not just Chinese, guanxi. In Hofstede's list, there were still several countries with individualism indices lower than Taiwan; most of them were in Latin America. From the guanxi gradient of Taiwan, it can be seen that the attitude of Taiwanese pilots toward acquaintances was extremely positive, the highest among all the pilots surveyed in this study. It is natural from the Chinese point of view. In reality, it is exactly the core of Chinese ethical value from Confucianism: close to the close. The acquaintances are very close to us, so we have to behave closely to them. Extrapolating further, if the data for kin can be obtained, the attitude would definitely be even higher for the Chinese. When the horizontal axis reaches the position of alien, the attitude of Taiwanese pilots was even negative. Actually, it is the Chinese tradition evolving from the long-lasting, highly developed Chinese civilization: carefully differentiate Chinese from barbarians. Taiwan, with such a strikingly steep gradient, is naturally not to be understood by the simple concept as a low individualistic country by Hofstede. Individualism, as generally accepted, represents a concept stressing human independence and the importance of individual self-reliance and liberty. However, for nations such as China with a low individualism index, some clarifications are clearly needed.

In Hofstede's work, and also as generally perceived, collectivism is a concept opposite to individualism. Therefore, for the countries with a low individualism score such as China, one would naturally assume that they must have a high degree of collectivism. However, is China a highly collectivistic country?

This point has already been denied by many Chinese scholars (Fei 1947; Hsu 1970; Yang, C.-F 1993; Yu 1984). It can only be said that China is a non-individualistic society. However, using the concept of non-individualistic to understand China is obviously not enough. In this regard the "guanxi gradient" can be treated as a complementary concept to individualism. Using individualism to describe interpersonal relationships for Western societies, including Australia, is appropriate without any doubt because the concept of individualism is derived from their society. However, since individualism and collectivism is a very rough dichotomy, non-individualism is often easily confused with collectivism. For a country such as China, interpersonal relations are surely non-individualistic but are also definitely not collectivistic. From the results in this study, it seems that individualism categorizes people simply into two classes, "self" and "others." As a result, the dichotomous view of individualism–collectivism is naturally formed. No further differentiation concerning others is made. As for a society like China, people further differentiate "others" into several categories (Jing 2004). There are many diverse "others" in the minds of Chinese people. Some "others" are even more different from other "others." The concept of collectivism is certainly not enough and sometimes is even misleading in describing the Chinese type of non-individualism. The concept of "guanxi gradient" is hence much more suitable to describe different types of non-individualism, including that of the Chinese. Of course, it is also appropriate to portray individualism by a flat guanxi gradient. Furthermore, the guanxi gradient is not a simple dichotomy. This gives the concept adequate scope to delineate different types of interpersonal relations, including individualism. In this aspect, the "guanxi gradient" is more useful than individualism and collectivism.

The primary goal of this study is to enhance global flight safety through promoting a mutual understanding between people with different cultural backgrounds, especially those who are involved right across the whole aviation environment. From the study of Helmreich and Merritt (1998), among the pilots from all over the world, the strongest differences were from command roles and interactions, attitude toward orders, rules, and routines, and preference for and reliance on automation. Although these items looked like describing something about authoritarianism, it was found that their correlation with power distance was only 0.73 while it was 0.83 with individualism. It implied that interpersonal relationships might be more fundamental in describing the global difference at national level. The statistical analysis of authoritarianism and flight safety had been executed (Jing et al. 2001) and a relatively high correlation was found, approximately about 60%. The correlation of power distance with difference for global pilots was over 70% and the correlation with individualism was even higher, over 80%. If the simple dichotomous individualism can have such a high correlation with the national difference among pilots, what could it be if the correlation was calculated with a more complete concept such as the guanxi gradient? It will definitely be more than interesting to execute this study. Therefore, the relationship between the guanxi gradient and flight safety was investigated.

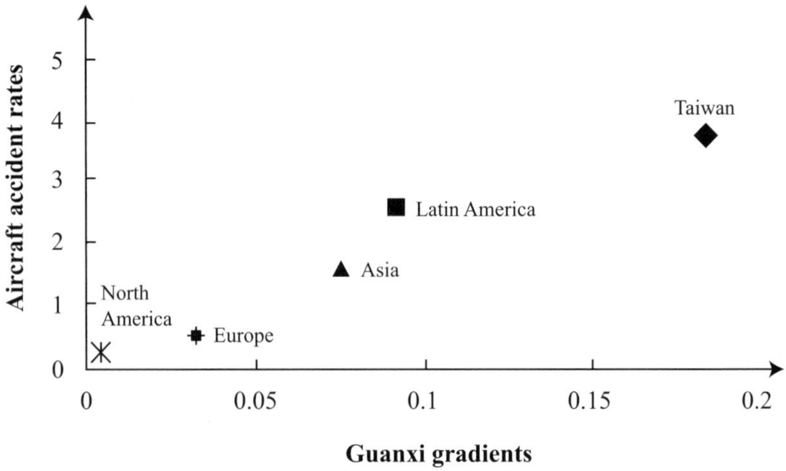

Figure 8.4 Global aircraft accident rates and guanxi gradients

The aircraft accident rates from different regions (Jing et al. 2001) were used as indicators of safety. The accident rates were plotted in relation to the guanxi gradient, as shown in Figure 8.4. For the initial study, only a correlation was calculated; however the correlation coefficient was found to be positive and surprisingly high at 0.977, a number that can almost be described with mathematical formulation. Of course, it will not be that simple.

A high correlation does not necessarily mean a causal relation exists. However, the correlation of safety with the management of interpersonal behavior is accepted by most in the aviation community and lies behind the concept of crew resource management. And, there were incidents caused by interpersonal relations in the cockpit being too close. On the other hand, we included in the sample 40 from North America, 21 from Latin America, 28 from Asia with Taiwan excluded, 32 from Europe, and 14 from Australia. Such a small sample is definitely not enough to make any concrete conclusions. Hence, the correlation of 0.977 cannot be treated as a rigorous finding. On the other hand, the chance of such a number arising purely by chance was also very unlikely. Therefore, the reasonable attitude would then be: how high the correlation is between the guanxi gradient and flight safety is still uncertain, but it is definitely not to be negligible.

With the results shown above, two things deserve attention. First, for global flight safety, the explanatory power of the differentiated order and the guanxi gradient seemed to be higher than individualism. Individualism is a kind of dichotomy based on the linear mode of thinking, and it cannot cover the whole spectrum of globally different interpersonal relations. The guanxi gradient was derived from the different Chinese ideographic mode of thinking and could have more chance to delineate human relations between those from different cultural backgrounds. Secondly, the correlation between interpersonal relations and flight safety does exist and the corresponding correlation with the guanxi gradient is so high that it cannot be disregarded. Further research is definitely required.

From the data, it seems that a country with a higher guanxi gradient will have higher accident rates. Superficially, as frequently discussed by scholars and journalists, the interpersonal relationship derived from the concept of family is "outdated" and "backward," even harmful, so as to result in such a high accident rate. Even if correct, this viewpoint is not helpful, because it is not possible for the Chinese to give up their culture, to cut off such a long-lasting tradition. Besides, this viewpoint is still debatable and not necessarily correct.

Aviation technologies including CRM are products of Western science, which means nothing in aviation technology is from other civilizations. Western science is also a product of linear thinking derived from the Greek alphabet (Jing 2012). When the aviation technologies were imported into an environment created by ideographic thinking, conflict is doomed and cannot be avoided. Adaptation is a must. The corresponding operations of CRM in the cockpit were designed by the individualistic culture, which does not value the differentiation of human relationships – the basic beliefs of which would naturally be high individualism and low power distance. From the perspective of the guanxi gradient, CRM is an institution with a flat guanxi gradient as its basic assumption. If it is operated by the crew with zero guanxi gradient, it would definitely be very smooth. However, if the crew possessed a guanxi gradient as high as 0.2, as with Taiwanese pilots, two institutions – one in the machine and the other in the crew – with such a great difference will clash in a small cockpit, leaving a very low chance of working smoothly. The conflict of the interaction between the crewmembers and the machine can certainly be expected. In this case, if something bad happened, it would be very difficult and unfair to blame one side only. Hence, in the high correlation of the guanxi gradient and accident rates, it would be not correct to talk about the pilot's error without referring to the machine, even when the machine works "normally." It is a problem of compatibility. Since the aviation system was designed by people with almost zero guanxi gradient, as the gradient of the crew started to increase, the incompatibility between them also began to grow along with the accident rates. Consequently, the extremely high correlation between the guanxi gradient and accident rates should be totally expectable and acceptable.

Cultural problems are indeed important in the field of global flight safety. However, to fix problems stemming from culture is extremely difficult, if not impossible, since it takes time for a culture to evolve and adapt. It is said (Harrison 2006) that cultural change is measured in centuries instead of decades. Actually, culture does change, but different cultural traits change at different speeds (anisotropy). This study provides some empirical results to support this argument. From the local Taiwanese results, the guanxi gradient was further broken down according to the age of the pilots, using five years as an interval to study the potential change in the guanxi gradient. The results are shown in Figure 8.5.

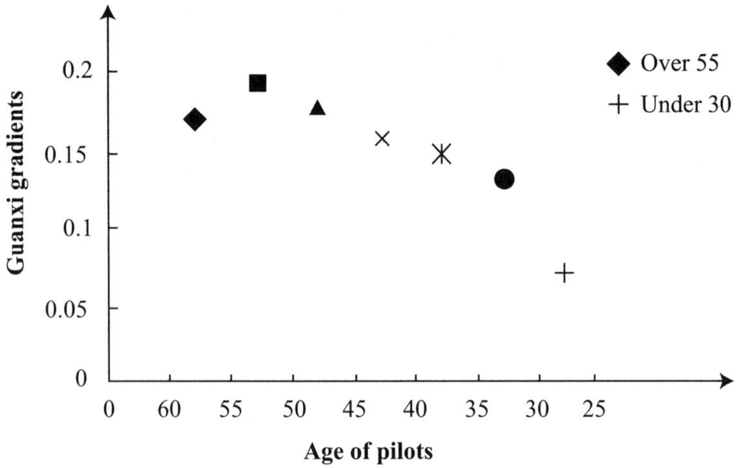

Figure 8.5 Guanxi gradients for different ages of Taiwanese pilots

It is found that generally the younger pilots possess a lower gradient, which means young people in Taiwan are more and more individualistic, since the gradient is flattening. This is quite reasonable. Taiwan has been Westernized and heavily influenced by the US for more than 60 years. Individualism has been accepted as the universal value. This tendency was again best fitted with a straight line representing the average decaying gradient to calculate the speed of change. Since the horizontal axis is time and the vertical is the guanxi gradient, the slope is the rate of the change in the gradient per unit of time. The slope was calculated to be less than 0.01. If the results are extrapolated up to a point that has the same gradient as that of Australians, it suggests that according to the present rate of decay, the time needed for the local pilots to have the same guanxi gradient as Australian pilots is calculated to be more than half a century. Since the data were taken in 2000, if the trend is kept unchanged, at least, after the year 2050, the interpersonal relations of Taiwanese society might very much be like that of Australia –assuming, of course, the situation in Australia remains unchanged.

Globalization is a hot topic worldwide. Following the diffusion of Western technologies and ideologies, the whole world is becoming more and more Westernized. Individualism is one example of this process. Figure 8.5 clearly depicts the evolution of the guanxi gradient in Taiwan. The flattening process is very obvious. Young pilots possess a flatter gradient. However, for the pilots over 55, the flattening gradient is also found. It is thought to be natural and can be explained as follows. For older people, the social pressure for them to adhere to the norms and values is lower because they are becoming more and more insignificant to the harmony of society. This is also acknowledged and accepted by Confucian teaching. What is interesting in this figure is the speed of the evolutionary process.

It is calculated to be more than half a century before the local pilots will likely have the same gradient as that of Australian pilots if the rate of change remains constant. Although Taiwan is already a very dynamic society, it will still take many years to evolve. An even longer time is needed for Chinese authoritarianism, which is estimated to be more than 200 years (Chang 2006). It was also reported in Hofstede's work (1980) that the answers to the questions about power distance did not shift over four years. However, the change in individualism scores is obviously more drastic. Both of these are consistent with the present study. For authoritarianism, the shift may be too slow to be seen in such a short time of only four years compared with 200 years. "Slowly" is undoubtedly the key word in cultural evolution. In addition to this, different cultural traits evolve at different speeds, so anisotropy is the other key trait that cannot be ignored.

Since the goal of CRM is to enhance cooperation in the cockpit, how performance is influenced by the attitudes related to intimacy–estrangement between the crewmembers is very important. Will some activities be distorted by the strength of these attitudes? As it was designed by Westerners, CRM conveys subconsciously the individualistic mentality. Some distortion in intent is therefore almost certain to take place in a cockpit with a steep guanxi gradient. By looking at Figure 8.3, the most serious differences take place at the two extremes with too much and too little intimacy, i.e., too good and too bad an attitude. To answer this question, in-depth interviews with the local Taiwanese pilots were performed. A total of nine pilots, five captains and four first officers, were interviewed. The discussions focused on CRM activities including monitoring, communication, teamwork and workload, all aspects that are possibly distorted by attitudes toward intimacy–estrangement. The local pilots were asked to supply concrete evidence from their experience to support their arguments. The results from interviewing the local pilots, summarized in Figure 8.6 (below), show the positive and negative influences. For closely intimate relationships in the cockpit, communication and cooperation will be enhanced. On the other hand, the monitoring function is likely to deteriorate. Work-sharing is found to be another function easily confused, because each crewmember is more than happy to help the other when they have a very intimate relationship. In a cockpit with estranged relationships, the bright side is that monitoring functions will be preserved almost as designed, and the procedures will be followed automatically because each crewmember does not want to cause any trouble. The downside of this, however, includes deterioration in communication and cooperation. Also, the degree of teamwork is highly likely to be lowered. Generally speaking, from this study it is found that in a cockpit with a steep guanxi gradient, the designed activities will be more easily influenced by personal feelings or emotions. In this case, self-discipline is needed to avoid any negative influences. However, this is not always easy for a less mature person and a good way to deal with this is to avoid a steep guanxi gradient in the cockpit, i.e., not to put two pilots with too intimate or too estranged relations together in the narrow and coercive space of the cockpit.

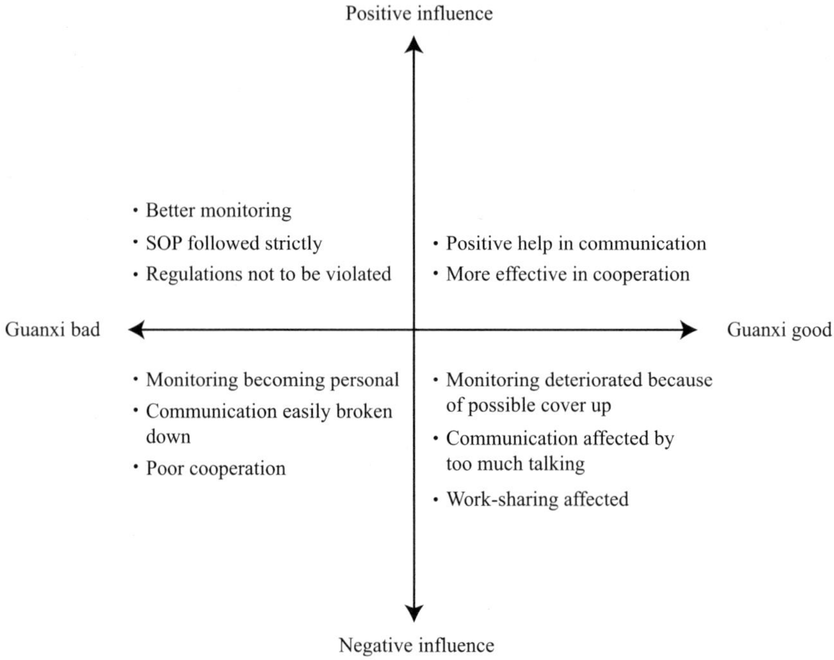

Figure 8.6 The role of guanxi gradient and its influence on CRM activities

With the goal of promoting global flight safety by enhancing the mutual understanding between people with different cultural backgrounds, the core concern of this study was to try to avoid the bias that may be encouraged by the dominant view of individualism and collectivism proposed by Hofstede. A new cultural variable, the "guanxi gradient," is proposed in this study. Based on numerical data, it can be seen that people do differentiate interpersonal relations and the gradient type of attitude does exist. Individualism represents one end of the spectrum and in this case differentiation only takes place between "self" and "others." No further differentiation exists. Even in this case, a gradient type of attitude still exists, since it will be very hard to believe that Americans treat their family members the same as other more distant nationalities. The difference is that Chinese, on the other side of the spectrum, further differentiate the "others" and use different attitudes to deal with different "others." The guanxi gradient can thus be used without difficulty to describe interpersonal relations for the whole spectrum. Consequently, the guanxi gradient provides a possible alternative viewpoint to individualism and collectivism to understand global interpersonal relations between different cultures.

With the surprisingly high correlation of 0.977, the relationship of the guanxi gradient to global flight safety cannot be neglected. However, accidents caused solely by a steep guanxi gradient are rare. What is really important is that the influence of the guanxi gradient will be magnified when it is intermingled with authoritarianism, as happened in the Songshan accident on Chinese New Year's Eve described in Chapter 2. The nature of the interaction between these two dimensions in the cockpit, superior–subordinate and intimacy–estrangement, is therefore the next important question to address.

The evolution of the guanxi gradient is clearly described in this study. However, cultural change is extremely slow. Further, different traits evolve at different speeds. Slowness and anisotropy are the two characteristics found in the evolution of culture. Patience in addition to respect is thus the necessary attitude when discussing cultural issues.

In-depth interviews with the local, Taiwanese pilots also revealed the influences of the guanxi gradient on resource management in the cockpit. Generally speaking, from this study it is found that in a cockpit with a steep guanxi gradient, the specified procedures will be more easily affected by personal feelings or emotion. Among the activities in the cockpit, monitoring and communication are the ones that are most easily distorted by the steep guanxi gradient. Given the extremely slow pace of cultural change, the appropriate way to deal with this will be to avoid any steep guanxi gradients in the cockpit by better crew assignment procedures.

Finally, one thing has to be stressed. The explanatory power of the guanxi gradient is not limited to flight safety only. Since the guanxi gradient is a variable describing interpersonal relations in a society, it can then be used in any field operated by that, which means almost everywhere in a society. There is a Corruption Perceptions Index (CPI) published by Transparency International. The Corruption Perceptions Index is defined as the ranks of the countries and territories based on how corrupt their public sector is perceived to be. It is a composite index – a combination of polls – drawing on corruption-related data collected by a variety of reputable institutions. The CPI claims to reflect the views of observers from around the world, including experts living and working in the countries and territories evaluated. There is another one called the Democracy Index (DI). The Democracy Index is an index compiled by the Economist Intelligence Unit, a private business based in the United Kingdom, that measures the state of democracy in 167 countries, of which 166 are sovereign states and 165 are United Nations member states. The index claims to be based on 60 indicators grouped in five different categories: electoral process and pluralism, civil liberties, functioning of government, political participation, and political culture.

The Corruption Perceptions Index and Democracy Index of the corresponding countries studied in this research were collected. After combining into an average number representing the data for that region – Taiwan was excluded from Asia due to the variation in the size of the samples – they were plotted together with the corresponding guanxi gradient as shown in Figure 8.7. The tendency was quite clear: the higher the guanxi gradient, the lower the Corruption Perceptions Index and Democracy Index, as expected. From the figure, according to the data, it is very clear that a society with a high guanxi gradient will be more corrupt and authoritarian.

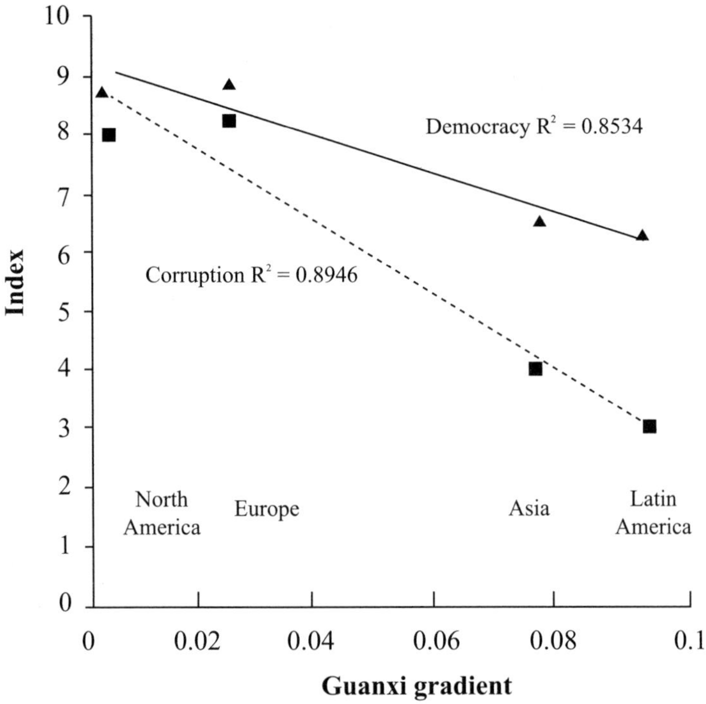

Figure 8.7 The correlations between the Corruption Perceptions Index and the Democracy Index with guanxi gradients

Those countries with highly democratic and non-corrupt public services are mostly highly individualistic. The correlation coefficients were also calculated, although the data were limited. The correlation between the guanxi gradient and CPI was found to be 0.89, and the correlation with DI was 0.85. Hence, it is reasonable to say that the Corruption Perceptions Index and Democracy Index are basically measuring the guanxi gradient. Again, it could be extrapolated as to say that the high guanxi gradient society is "backward" and "corrupt." However, it is obviously misleading. We have to be very careful of one thing here: this phenomenon really implies incompatibility only. Based on high individualism and low power distance, the Western legal and political systems are highly culturally specific. And it should not be used as the sole standard to measure other cultures, not to say providing any judgments about other cultures, since it will be very strange to say individualism is advanced and non-corrupt, and non-individualism is outdated and corrupt.

Chapter 9
A Struggle of Two Mindsets

9.1 The Lack of Completeness of Linear Thinking

Western societies have long been using an alphabetic language and writing, forming the kind of thinking that emphasizes sequence and logic. Within this mindset, what would they see and what kind of perspective would they hold when facing aviation safety? What we can be sure of is that they will undoubtedly see a sequence that is lined up according to the logic of cause and effect. And if they want to find some everyday words to illustrate aviation safety, the domino, the chain, or James Reason's "swiss cheese" model (emphasizing defense in depth) would appear to be very natural and even logical choices.

However, this mindset has a very critical and obvious flaw. Sequential, "logical" thinking comes from an alphabetic language, and it confines people's mindsets to be a linear one. This type of language relies on sounds to convey message, and sounds come from the vibration of air, and the vibration is actually the variation of the air density along with the time. Without other tools such as facial expressions, gestures, or graphics, this form of communication can only be one-dimensional. One need only compare a telephone conversation (in which words, tone of voice, and phrasing are the only communicative resources) with a face-to-face conversation (where facial expressions, body posture, gestures, ambient surroundings, and proxemics are also part of the communication) to appreciate this difference. That is to say, the alphabetical language is like a one-dimensional filter that channels the conveyed information into a sequence of unitary signals aligned in a "logical" order. So it is almost inevitable for Western thought to be confined to a linear channel, ever since the creation of the first alphabets more than 2,000 years ago – the first complete thinking tool ever created in Western history (Fischer 2001; Logan 1986; Man 2000).

Alongside the Western mindset being linear, when facing an aviation accident, this style of thinking would continuously build up sequential flight safety theories focusing on cause-and-effect relationships. However, the thinking can be linear without any difficulty, but not the actual facts. So if the facts are also being pushed into a linear pattern, omissions will necessarily happen.

There are three focal points for an aviation accident investigation: findings, cause, and recommendations (Wood and Sweginnis 1995). Supposedly, findings are those factors which are significant to the accident. Causes are those findings which are causal, and the recommendations are the suggestions of the investigation body as to what should be done to prevent recurrence of the accident. After all significant conditions and events, causal and non-causal, are found in the investigation, they are arranged in the order in which they occurred. The findings then become a sequence of events. Obviously, we look for findings so that the cause of an accident can be found in order to prevent it from happening again. Hence, cause identification is really at the core of the investigation.

However, there are so many different kinds of causes, e.g., primary, contributing, main, direct, underlying, root, probable, related, proximate, and even aggravating causes. There is a formal definition by the ICAO, which is the actions, omissions, events, conditions, or a combination thereof, which led to the accident or incident. The National Transportation Safety Board of the United States uses "probable cause" and accepts it as the description of the accident event along with its cause-and-effect relationship. The US Air Force had a clear description for the cause to be an act, an omission, a condition, or a circumstance, and it either starts or sustains the mishap sequence. From the point of view of an outsider, it seems that everybody agrees with arranging the findings in a certain order or sequence of a cause-and-effect relationship. This perspective looks perfectly fine on the surface, because that was how the accident developed. However, people do have different views on how the definition of the cause should be given and whether it is logical to explain how the accident happened with these causes.

In fact, the real problem isn't how to define the cause; the real problem actually is the inclination of causing misunderstanding when using a sequential cause-and-effect perspective trying to explain the whole picture of the accident. And it just makes everyone believe that the accident really happened in a sequential way, so different opinions arise about which and what the cause should be. Using linear thinking to explain how an aviation accident happens is like drawing a line curving through a painting and showing the audience how the painting should be interpreted and understood. If we look at the painting following the line, we actually find that we've lost most of its information. No matter how complete this line is, it would still be very difficult to grasp the whole picture of this painting, and people would always be discussing which should be on the line which should not.

Turkish TK981 Accident

Aviation is a very complex system. There could be flaws anywhere inside the system. And once there's one mistake in the system, very likely it will just initiate an unexpected chain reaction in multiple directions. Therefore, any flaw or mistake in the system should also be viewed in a systemic angle.

On March 3, 1974, Turkish Airlines flight TK981 planned to fly from Paris to London. This plane was a DC-10 with three engines, two below the wing and one below the vertical stabilizer. It was 11:00am; the plane carried 167 passengers from Istanbul, and landed in Paris. Of the 167 passengers, 50 got off the plane because they were travelling only to Paris. Then, another 216 passengers boarded the plane and prepared to fly to London. At 12:20pm, the plane was ready to take off.

During the take-off and climb, everything was normal, and at 12:37pm the crew reported an altitude of 7,000 feet and continued ascending.

Four minutes later, the cockpit voice recorder recorded the crewmembers talking in Turkish in a very pressing voice, with clear and loud background noise. The conversation and background noise lasted for about 30 seconds, and then the cabin pressure warning device started ringing, which showed that the cabin had lost pressure. Then, the over-speed warning rang too. Very soon, the radar screen showed that the plane started to bank leftward and descended at a very high speed. After no more than a half minute the plane disappeared from radar. Soon afterwards they found the plane crashed 40km northeast of Paris. Including the crew and passengers, 346 people died. No one survived the accident.

In the investigation, people initially suspected that the accident was caused by a terrorist attack. Since they found six bodies and parts of the wreckage of the plane 15km away from the crash site, many people believed that the plane exploded midair. Furthermore, after a thorough review of the radar screen record by the investigation team, they found that just at the time when it showed the clear and loud background noise inside the cabin, the plane, a light spot on the radar screen, separated into two parts. The bigger one went off leftwards on the radar screen, and the smaller one stayed at the original spot, and disappeared after 2 minutes. Obviously, when the plane was still in the sky, it was no longer one complete body. However, from the post-mortem tests of the six dead bodies, they found that they all died from a huge force, not of anything related to an explosion. Even when they looked into the six seats falling together with them, and the floor in the cabin, they couldn't found any evidence showing that there was an explosion.

After a prolonged investigation, the truth was gradually revealed. The cargo door to the rear port side of the plane was torn off during the flight. And this was because of a flaw in the design of the cargo door. After the crew on the ground closed the door, it was required to lock the door manually with a 30-pound force. And if the latch isn't locked properly, the air pressure in the cargo compartment would gradually shift from the latch to the bolt of the actuator while the plane continues ascending. Yet this bolt isn't designed to resist the cargo cabin pressure. Therefore, the air pressure could actually shear the bolt when the plane flies up to about 10,000 feet.

When TK981 ascended to around 9,000 feet, the bolt of the actuator finally sheared under the growing air pressure, and at the same time, the cargo door exploded away together with parts of the skin, causing the cargo compartment to depressurize. And the cargo compartment is immediately beneath the passenger cabin, separated only by a thin floor with some supporting beams. At the time when the pressure of the cargo compartment suddenly went out, the cabin floor buckled, since it was under extremely uneven pressure from two sides. It tore apart parts of the cabin floor, and two rows of seats together with six passengers were sucked out the plane. Yet, what happened afterwards was even more fatal.

In 1974, the fly-by-wire technology was not in place; the flight control system was designed using mechanical steel cables installed inside the plane, either inside the walls or under the floors, to convey the order from pilots to the control surfaces.

And while TK981's floor near the tail of the plane broke, part of the control cables that go through under the floor were torn or snapped or severely jammed. And of the four steel cables that control the elevators, the two that control climbing were torn down and the two that control descent were pulled tight, causing the nose to dive rapidly, and at the same time preventing the pilots from pulling up the nose.

Simultaneously, the cables pulling the rudder to the right were torn and the cables pulling the rudder to the left were pulled tight, causing the plane to turn to the left. With the one engine under the vertical stabilizer no longer functioning (after the floor in the tail broke), the pilots completely lost control. They could not adjust the thrust from the engine, and couldn't even shut it down. Several fatal factors happened at the same time, and the airplane dived and turned leftwards with an increasing speed. When the captain shouted "Bring it up, bring her nose up," the first officers replied, "She's not responding." Finally, TK981 hit the ground at a speed of 800km per hour.

According to the sequential cause-and-effect perspective, the accident chain of TK981 can be described as in Figure 9.1. This sequence is basically correct with what has happened. It's just that a linear explanation cannot capture the whole picture of this complex cause and effect in such a complex situation. The major issue is this: the sequential explanation will mislead people to think that there was only one thing that happened at each step. And each event in between is the effect of the previous one and also the cause of the next one, as given in the theories such as domino and accident chain. However, if we see it from a systemic angle, we would find that at any moment, there were many different things happening, not just one. And each event in between is the effect of many factors that happened before, and also the cause of many factors that followed.

Complex Systems

With technologies developed on an industrial scale, safety becomes a critical issue. Some technologies operate in extreme environments, or handle dangerous materials, or large amounts of energy. Once the operation does not function properly, the loss of control can result in a minor incident, a major incident, or an accident.

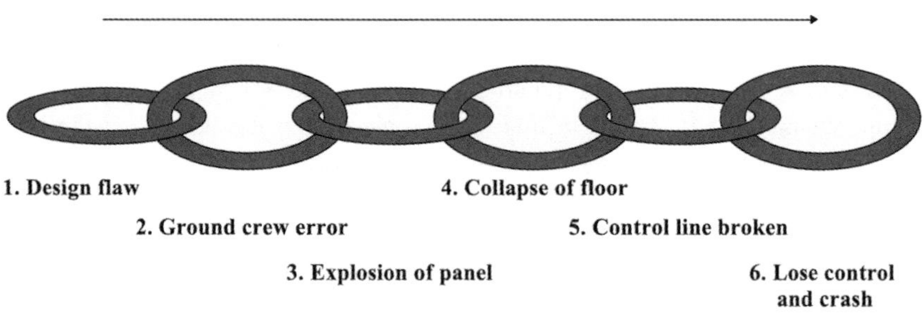

Figure 9.1 The accident chain of the Turkish TK981 accident

Some accidents, such as the Fukushima or Chernobyl reactor accidents, become disasters by shaking the foundations of the entire industry or society. These are high-risk technologies (Perrow 1984). They include nuclear power plants, chemical plants, and civil aviation. Once there is an accident, no matter whether it is radiation leakage at a nuclear power plant, explosion of a chemical plant, or a plane crash, it will always cause severe damage to human lives and properties, and even possibly affect the entire development of the society.

There are some common traits for thess high-risk technologies:

1. operation with dangerous substances;
2. control of large amounts of energy;
3. operation in extreme environments;
4. large groups of people affected.

In order to better operate these high-risk technologies, people use even more advanced technologies to control it. And this caused the operating system to be even more complex and complicated day after day. The more complex the system is, the more types of accident it can trigger. The more complicated the system is, the more chances there are for the different sub-systems to be coupled in complex ways. Therefore, when an accident happens, the possibilities of all kinds of different interactions between sub-systems become a lot higher. A very minor mistake will then trigger an unexpected chain reaction and cause an unthinkable result which actually brings huge disaster – such as in the Nagoya accident, which happened only because the co-pilot inadvertently triggered the go-around lever and caused a chain reaction that then brought the unexpected interaction between the sub-systems, leading to a huge tragedy.

In high-risk technology, Perrow described the two main traits as the reasons why high-risk technology is prone to lead to unexpected accidents. The two traits are high complexity and tight coupling. Any system with these two characteristics is sufficiently prone to failure that such accidents, according to Perrow, should be considered "normal." Civil aviation is a typical complex system. In a complex system, there exists a very high potential of unknown interaction between sub-systems and elements. The most distinct feature of the complex system is that any minor mistake will induce an unexpected chain reaction and cause incredibly severe event. This nonlinearity is a characteristic of complex systems. From a systemic angle, the fundamental reason why any minor mistake can trigger an unexpected chain reaction is because that at any single second, there are simply many things going on simultaneously in the system, and they are also interconnected.

For example, when the cargo compartment door of TK981 broke because of the poor design and poor maintenance from the ground crew, the following things actually happened at the same time:

- the cabin pressure decreased;
- passengers and crewmembers suffered from loss of oxygen;
- the cabin floor structure buckled and the function was broken;
- the structure of the plane deformed;
- the aerodynamic shape of the plane changed;
- the structure of the plane started shaking;
- the control cables could not function properly;
- the tail engine lost control.

All of the above events happened together and were continuously evolving while at the same time affecting each other. The plane crashed because it lost control. Ultimately, however, this did not mean the other parts of the plane functioned normally. The reality is: all other parts of plane were deteriorating; it's just that the single event "the control cable lost function" evolved faster than the others.

If the crewmembers were highly capable and could even just barely control the plane to continue flying, the single event "the control cable lost function" might have evolved slower or even been tamed. Could the plane possibly be safe under such conditions? Could they prevent the plane crash? Even had the plane kept on flying by superb airmanship, the cabin suffered a shortage of oxygen, which might have caused a crewmember to lose consciousness and crash the plane. Or even if the crew could maintain consciousness with the shortage of oxygen and keep the plane in the air, the structural damage might become more and more severe because of the vibration, the tail wing could possibly be torn off because of the high stress induced, and then caused the plane to crash. Or … there were simply just too many other possible routes for events to proceed along that might bring the plane to crash. This is characteristic of a complex system.

With such a highly complex and tightly coupled system, it strongly suggests that the linear mode of thinking emphasizing sequence could not delineate the complete picture of the complicated and multiple causal relationships in flight safety. In a complex system, any single event will have multiple effects on the system and form a complex situation where multiple events are happening all together at the same time. The evolution of these events actually forms a tree with interlocking branches. And which event will eventually lead to the crash of the system is highly uncertain. Even when one of the events finally evolves into system breakdown, the event is still affected by many other events. Typically, the investigation of a flight accident starts from the final outcome tracing back upstream to the first cause following the cause-and-effect relation. After investigation, by placing all the events according to the cause-and-effect relation, it would be very natural for people to have a sequential description about the cause of the accident, just like the domino theory, the accident chain theory and the defense-in-depth theory. However, there are still many other events affecting and contributing to the evolution process of the accident which are not directly located in the sequence. They are situated in the multiple cause-and-effect relations but not involved directly in the sequence leading to the occurrence of the accident.

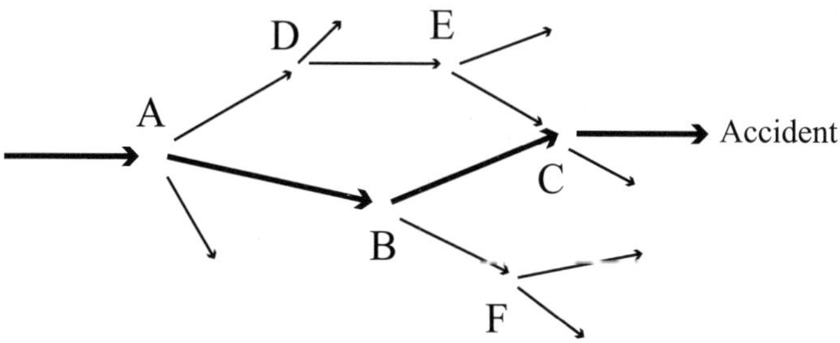

Figure 9.2 Comparison between a sequential cause-and-effect perspective and a process of how the accident developed in a complex system

That is thought to be the reason why there are so many different kinds of definitions for the causal factors in the literature.

Now, let's use a simple schematic diagram to show the deficiency of linear thinking in flight safety. In Figure 9.2, through the investigation of a flight accident, there are many findings identified, e.g., A, B, C, D, E, and F. They could be the mechanical, environmental, or human factors. After carefully reviewing those findings, it is concluded that factor C is the direct cause of the accident. Then the investigation further finds that factor C was actually initiated by factor B, and factor B was again initiated by factor A, and A is found to be the first cause of this accident chain. If a sequential causal relation perspective is used, the sequence of A-B-C-accident will then be the conclusion of the investigation process of this accident, along with recommendations to prevent A, B, and C from happening again.

However, if we look at the complex causal relations in the evolution process in detail, we will find that factor A not only initiated factor B, it also initiated factors D and E. In addition, factor B not only initiated factor C, it also initiated factor F. Yet factors D, E, and F won't appear in the causal chain in the investigation report since they are not the factors leading to the accident as defined by the ICAO. They will also be rejected by the US Air Force, since they are not the factors in the mishap sequence. But the influence of factors D, E, and F is still there. They have also contributed to the whole process of the accident. It's just that these factors do not directly lead to the occurrence of the accident; it may be just because they evolved more slowly, or they were confined or contained. When we use a traditional angle to view causes in flight accidents, they will not be counted as a cause. But they do contribute to the accident, sometimes in a very strong way. Thus, it will surely arouse controversy if we take them away from the causal chain.

The deficiency of a sequential or linear cause-and-effect perspective not only arouses controversy. It might possibly cost us some insight into flight safety under certain conditions. A linear mindset suggests that flight accidents happen because a chain of events with cause-and-effect relationships took place under certain specific conditions and linked together. Based on this type of thinking, anyone would conclude very naturally that if one of the links is taken out, the chain will be broken and the causal relationship will be cut off. Hence, the accident will be stopped from happening. So, everyone comes to believe that to stop similar accidents from happening again, all we have to do is break the chain.

According to the investigation report, the whole process of this certain accident is from A to B to C and then to the outcome. Therefore, to prevent similar accidents from happening again, we could just break the chain at any point, e.g., between B and C. If the link between B and C is taken away, it means even if event B took place, it would not induce C because there is no route to induce C from B. So, the airplane manufacturers then took out the link from B to C by modifying its design, so that even when B happened, it would not lead to C and a similar accident will be prevented. However, an airplane is a complex system: every event will induce multiple effects because of the highly complex and tightly coupled nature. When we look back from the outcome to the first cause from a causal relation point of view, we will see a linear process. However, if we look from the first cause to the outcome, we will see a different picture because every cause will induce multiple events. It will then look like an event tree, not just a line, and there might even be complicated linking between those branches.

Therefore, the next time similar conditions appear, event A materializes, and after B is induced, the chain is stopped because there is no route for B to induce C. Unfortunately, in addition to B, event A also induced D because of the tightly coupled nature of a complex system. After D, E is triggered by D and then E will very likely cause C to happen. If that is the case, there is a very high probability that the same accident will happen again. Therefore, even after we take out the link between B and C, and prevent the accident from happening through the route of A to B to C, the accident might develop through route A to D to E to C, and cause exactly the same accident, just as Figure 9.3 suggests. Hence, the belief that a similar accident can be prevented from happening again by breaking the chain is not correct. There are simply too many causal routes in a complex system.

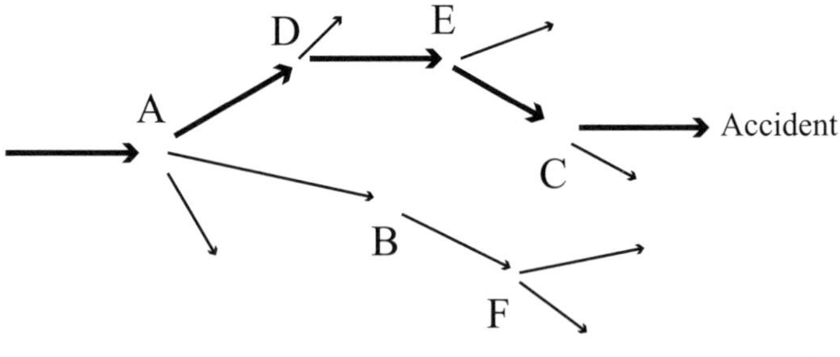

Figure 9.3 In a complex system, a similar accident may develop through the route A-D-E-C, even if the link between B and C is taken away

One chain is broken, and another chain will be formed. We cannot be certain that we will stop a similar accident after we take out the link somewhere, because it might develop itself through some other routes. To put it in plain language: an accident will find its own way out.

From Nagoya to Da-Yuang

This hypothesis doesn't just come from pure imagination. The occurrence of the Da-Yuang accident after Nagoya provides confirmation.

The core cause of the Nagoya accident is thought to be the conflict between the pilots and the machine, or aircraft–pilot coupling (National Research Council 1997). The first cause in the accident chain was that the first officer inadvertently triggered the go-around lever, causing the automatic flight system to execute a go-around, yet the crew still wanted to land. Based on the Airbus design, when the automatic control and manual control are in conflict, the trimmable horizontal stabilizer and elevator would clash. In the Nagoya accident, the abnormal out-of-trim condition continued to eventually trigger the alpha-floor protection. The angle of attack increased due to the large pitch-up movement coming from the full thrust until the plane stalled and crashed to the ground. Stated simply, the Nagoya accident resulted from the fight for control of the airplane between the pilots and the automatic flight system. After the Nagoya accident, Airbus decided to yield the final authority over the plane in urgent situations back to the pilots to avoid such a conflict between the pilots and the electronic crewmember from happening. Airbus changed the design of the flight control system. In other words, as long as the crewmembers pushed the control yoke down with more than 33 pounds of force, the autopilot would be overridden, and the control over the airplane would be back in the hands of the crew.

However, in less than four years, an accident very similar to Nagoya, the Da-Yuang accident, happened. The cause of this accident was that the altitude of the plane was too high during approach. The plane continued to fly horizontally for about two minutes when it should have started descending at 29,000 feet. Facing such a scenario, the captain kept on lowering the flaps and landing gear, hoping to reduce the altitude by increasing drag. The altitude was depleted as expected. Unfortunately, before their action could reduce the altitude, the speed of the plane was too low and the angle of attack too high. This activated the alpha-floor protection function, and the pitch of the plane started to increase rapidly. At this moment, the pilot instinctively pushed the yoke down to lower the nose. However, since the pilot pushed it with more than 33 pounds' force, the manual control overrode the automatic control; the automatic control of the plane was deactivated. From the final records of the plane, it seemed that the crew did not know that the automatic flight system had been disengaged. When the captain finally decided to go around, the crew let go of the yoke for the automatic flight system to take over and execute a go-around. Yet the automatic flight system had already been disengaged by the crew themselves, leading to an 11-second period of no control at the most crucial moment, until the plane stalled and subsequently crashed.

After the Nagoya accident, theoretically, when we took out one link of the human–machine conflict and broke the accident chain by changing the design of the final authority over the plane, we should have been able to prevent similar accidents from happening again. However, that is not what happened. After the Nagoya accident, the automatic flying control system has been modified to avoid the fight over control with pilots. Yet, it was for this reason that the plane was not piloted at the crucial moment, and actually resulted in an almost identical accident in the Da-Yuang tragedy. Hence, when we thought we could rule out the possibility of similar events from happening again by taking out one link in the chain, the accident still happened by evolving through another route. The sequential causal relation-type of thinking is insufficient and incomplete when explaining flight accidents.

Risk Assessment of Flight Operations

The deficiency of linear thinking is clear not only in the prevention of accidents, but also in the risk assessment of flight operations. Why so? Basically, from a Western perspective we see the issues of aviation safety from a sequential cause-and-effect perspective. So, from an active accident prevention point of view, we want to break the accident chain, i.e., take out one link to prevent similar accidents from happening again, to simply "break the bad chain." But, this type of thinking was proven to be insufficient from the examples of the Nagoya and Da-Yuang accidents. On the other hand, from an aviation safety enhancement perspective, we would naturally want to maintain a healthy causal sequence, in other words, to keep everything following the standard operating procedures, so that we can lower the risks associated with flight by preventing deviation away from the standard normal flight. However, even if we want to break the bad chain or protect the good chain, we will have to see the aviation safety issues as a chain first, i.e., from a sequential point of view. Unfortunately, once flight safety is portrayed as a sequence, we are drawing a line across a complex picture. Lack of completeness is then inevitable.

In March 2007, based on the contents of Federal Aviation Administration AC120-92 from the United States (Federal Aviation Administration 2006), the Taiwanese Civil Aviation Authority issued a circular about the safety management system to provide guidance for local airlines to build a flight safety management program. In the circular, it clearly pointed out that the objective of the safety management system is to provide a structured management system to control the risk of flight operation. No matter how complex the entire aviation operation is, the core of this system is undoubtedly risk management. The safety risk management process is used to examine the operational functions of the company and their operational environment to identify hazards and to analyze associated risk. The main tasks of risk management include the following (Civil Aeronautics Administration 2007):

1. **System and task analysis**: Safety risk management begins with system design. The system or task descriptions should completely explain the interactions among the hardware, software, people, and environment that make up the system in sufficient detail to identify hazards and perform risk analysis.
2. **Hazard identification**: Hazards in the system and its operating environment must be identified, documented, and controlled. It also requires that the analysis process used to define hazards consider all components of the system.
3. **Risk analysis and assessment**: The standard risk analysis and risk assessment clauses use a conventional breakdown of risk by its two components: likelihood of occurrence of an injurious mishap and severity of the mishap related to an identified hazard, should it occur. A common tool for risk decision-making and acceptance is a risk matrix such as the one in the ICAO Safety Management Manual. The definitions of each level of severity and likelihood will be defined in terms that are realistic for the operational environment.
4. **Risk acceptance**: In the development of its risk assessment criteria, aviation service providers are expected to develop risk acceptance procedures, including acceptance criteria and designation of authority and responsibility for risk management decision-making.

In addition to the ones listed above, there are other elements to risk management. However, from the above description, we can clearly see that the crucial point in the flight safety management system is really the identification, assessment, and evaluation of risk. Among the seven phases of flight – from taxi, takeoff, climb, cruise, descent, approach, and landing – the most dangerous phases, with the largest percentage of accidents, are approach and landing. According to the definition of risk, and taking into account the likelihood and severity of the accident, the approach and landing phases are the ones with the highest risk. Therefore, approach and landing are also the focus of flight risk assessment and evaluation.

The giant civil airliner nowadays is already very advanced. Not only has it been highly automated, it also has systems to monitor the airworthiness of the plane in great detail. Today, when commercial aircraft are in operation, through the sensors installed in numerous locations in the plane, the entire flying situation is under constant surveillance and recording. For every flight, hundreds of operational data points such as altitude, speed, angle of attack, and many others, are recorded, providing very thorough information about the whole flight situation, especially approach and landing. All those recorded flight operational parameters can be used for the scientific and quantified evaluation of risk assessment of flight safety management. This is the Flight Operation Quality Assurance (FOQA) system.

The FOQA system can record hundreds of relevant data during flight. The manager can set standards to decide whether the flight operation quality is good or bad from these parameters. For example, in the final approach, the airplane should fly at a certain descent rate. It represents that the plane is getting close to the earth at a proper speed to insure a smooth landing. If the descent rate is too high, it might cause a heavy landing of the plane. Not only will the passengers be scared, but also the structure of the plane might well be damaged. If the descent rate is severely high, we might call it a crash. On the other hand, if the descent rate is too low, the risk of long flare or overshooting the runway will be increased. It is also very dangerous for landing. Therefore, if we can monitor the descent rate during approach, it will surely be helpful for flight safety. To do this, we need a threshold for the descent rate to keep the flight operation within a proper range around a given standard, which means we are actually monitoring the deviation from a standard. If all the parameters of flight operation in all phases can be kept within a small range of deviation from the standard, the flight is very safe. Further, the smaller the deviation is, the safer the flight could be. Hence, controlling flight operational quality through monitoring these recorded data is actually an ideal way to manage flight safety, and it has already been proven to be effective.

Using the FOQA system as a tool to evaluate and control flight operational risks, despite its obvious effectiveness, we still need to acknowledge clearly that the entire system is designed from a Western point of view, and will definitely be affected by the linear mode of thinking. Accepting that the flight operation process is a sequence, there will certainly be a good sequence formed by the standard operating procedures for every phase of flight. Every maneuver during flight has to comply with these SOPs. There will then be a standard series of numbers for each recorded parameter by the FOQA system. If each FOQA parameter can be kept very close to the standard sequence, the flight will certainly be very safe. Hence, such a good sequence has to be protected, and pilots are expected to strictly follow it. If the flight situation begins to deviate from it, the risk will be increased. By setting a threshold for every FOQA parameter, i.e., to set an envelope for each FOQA parameter sequence, the corresponding risk of flight operation can thus be assessed. With this, the flight operational risk analysis and assessment can proceed. As in Figure 9.4, if we set up a series of envelopes for these flight operational parameters to protect the good sequence of flight, and make the plane always fly within a small range of deviations from the SOP, the risk of flight is then controlled within a certain acceptance level. Take descent rate as an example again: there is always a standard sequence of descent rate from the SOP for the flight operation. When the descent rate is higher or lower than standard, deviation appears and the risk is increased. If the deviation is larger than a certain threshold, the risk can be set according to its severity level. If the deviation of the descent rate is even higher, the severity is promoted to another level, so as the risk. Hence, the management can set up a standardized level of severity as well as risk from the deviation by setting the threshold, such as low, high, or alert, warning, etc.

To protect the good sequence of flight is very good for flight safety. It is just not sufficient. It lacks completeness. The basic motivation of this technique is defining risk from the deviation of flight operation from the standard sequence.

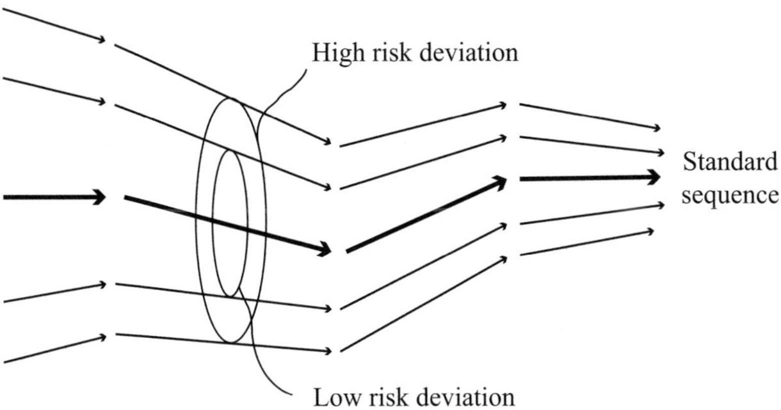

Figure 9.4 Safety management with Flight Operation Quality Assurance system using the basic concept of protecting the good sequence

The most fundamental problem of this linear mode of thinking is always starting from some kind of sequence. After that, we have standard sequence, and then we have deviation from the standard, and then we have severity and risk associated with it. But drawing a line across a picture to delineate the meaning of the picture is always insufficient. Take the descent rate as the example again. There is one scenario, A, in which the deviation of descent rate is smaller, but the altitude of the plane is also lower. The other scenario is B. In scenario B, the deviation of descent rate is larger, but the corresponding altitude is also much higher than scenario A. According to the normal Flight Operation Quality Assurance standard, scenario B is much more severe than scenario A. The empirical reality, however, could be just the opposite. A 600-feet-per-minute descent rate at an altitude of 200 feet is much more dangerous than a descent rate of 800 feet per minute at an altitude of 2,000 feet. It will be agreed by most that the same amount of deviation of descent rate occurring at different altitudes can have different severity for flight safety in general.

Since aviation safety is a complex system, there are many events ongoing at each moment and the interactions between them should not be neglected because of the tightly coupled nature of the system. As a result, the risk of every situation of flight should be evaluated from a systemic point of view. It means that the risk or severity should be a function of flight situation instead of deviation from a single parameter. Now, let us take Figure 9.5 for discussion. This figure shows a well-defined standard operating procedure with associated parameters of flight. By using the flight safety management system with FOQA, the flight risk assessment is executed by identifying the deviation of those parameters associated to the SOP. That is to say, the severity or risk is the function of individual parameter deviation. From Figure 9.5, if flight safety is viewed as a sequence, this kind of assessment is relatively correct.

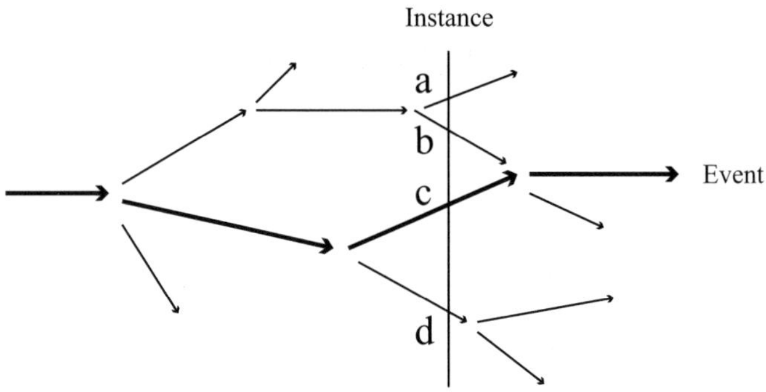

Risk = function of c (linear thinking)
Risk = function of a, b, c, d (ideographic thinking)

Figure 9.5 Lack of completeness in flight operation risk assessment using the linear mode of thinking

However, at any instance of time, there are so many parameters evolving together and interacting with each other and the risk or severity should be evaluated as a function of the complete flight situation, i.e., all the parameters defining the situation should be included interactively for the assessment to be closer to the reality. Therefore, according to the current flight safety assessment based on linear thinking, judging from the estimation of the severity of the flight operation by individual parameter deviation, it might well be proven from the above discussions to be a lack of completeness, although it is not wrong.

9.2 The Difficulty of Incorporating Ideographic Thinking in Aviation

Chinese ideographic thinking originates from the Chinese writing characters. This type of thinking is not from language, as in the West. It is a graphical type of thinking, or two-dimensional thinking. It has special traits that emphasize totality, treating the subject and object of thinking altogether, and equilibrium between components, with the advantage of insight to grasp a key feature with a small amount of information and identify the shortcut to find a way out. This type of thinking would unavoidably encounter very subtle incompatibility when facing an aviation system that is completely built on linear thinking. And the most serious incompatibility is that the linear view emphasizes sequence, but the Chinese do not, since sequence cannot be uniquely defined two-dimensionally.

As previously mentioned, Boeing conducted a detailed analysis on human errors by further differentiating human errors into 15 actual observable behaviors (Boeing Commercial Airplane Group 1994). Among all the aviation accidents around the world, not adhering to procedures as a human error showed up in almost half of the accidents. As shown in Figure 4.5, the average percentage of Asian pilots not strictly adhering to the SOP is 11% higher than that in the United States and Canada, and 14% higher than in Europe. These concrete numbers are thought to be in agreement with the general perception. Superficially, it seems that these numbers tell us that pilots from Asia have made more mistakes or are more negligent than their Western peers when it comes to abiding by the standard operating procedure.

Western languages are alphabetic. Multiple written characters are used to form words, which in sequence create sentences. Sequential thinking is an expectable (though not inevitable) consequence of having an alphabetic language. Therefore, it is normal for Westerners to abide by procedures, since this is consistent with the structure of their language. Following standard procedures is consistent with having a sequential language. By contrast, Chinese language does not have an alphabet, and Chinese writing basically is not related to pronunciation. That is to say, the most fundamental difference is that the Chinese character does *not* record Chinese language (Lee 2008). It is called the "separation of language and writing," i.e., what you see and what you read have almost no apparent relationship. Although there is an order when writing each stroke of a Chinese character, it is only a general guiding principle. It is totally acceptable for Chinese to write a word without going through the normal procedure; as long as it has the same features when it is finished, it conveys exactly the same meaning and can be used for communication. Moreover, the meaning of any Chinese character actually heavily depends on the context. And when some Chinese characters are put together to form a sentence, even the punctuation can completely reverse the meaning of the sentence if put differently. Therefore, the thinking, writing and reading mode of the Chinese language is definitely inclined to a graphical one.

Since the Chinese rarely need to go through the right sequence – they are very flexible with it – it is natural for Chinese people to seek shortcuts in whatever they see or hear, to find the meaning behind it in the fastest possible way. This is unlike Western language, which emphasizes sequence, procedure, routine, and regulation. It is because of this fundamental fact that the Chinese adhere less to sequence, procedure, and regulation when compared with Westerners. Given this, when Chinese pilots have to complete an assignment according to a very strict procedure, cultural pressure will build up immediately and the result will very likely be deteriorated, since the pilots will have to carefully concentrate on the procedures for them to be strictly followed. It means the pilots have to fight their own instinct. When we use the phrase "human error" to describe any incident resulting from this kind of incompatibility, we have actually mistaken a cultural phenomenon for an individual action, and this misleads us.

When a crewmember with an ideographic thinking mindset faces a flight environment built with a linear mode of thinking, cultural pressure is inevitable. The question is, how big is the pressure? And how would it affect the crewmember's performance? The cultural pressure caused by the difference between thinking modes includes many different aspects. For the sake of continuing this discussion, let's simplify the issue. We will focus on how big the effect can be from the difference of understanding of the standard operating procedure based on different modes of thinking.

Research on the effect of the fundamental difference in thinking modes on understanding the flight operation manual has been done in the Institute of Civil Aviation, National Cheng Kung University (Jing and Lee 2001). The results showed that a manual which is written totally based on linear thinking actually created a barrier to understanding for people who think in an ideographic way. If the difference of reading and writing habits from the modes of thinking can create a barrier to understanding the flight manual, it of course can affect how well the pilots absorb the content from the manual, and further affect the willingness of the pilots to strictly adhere to it.

Currently, the flight operation manuals are all written in English. Some airlines may use a translated version. But, in order to avoid unnecessary confusion or even misunderstanding, most airlines require all pilots to use the original English version. Even if there is a Chinese version, it is still directly translated from English in order to keep the information correct without any distortion. However, the fundamental characteristics of these two languages differ markedly in writing and reading, even in thinking. Westerners are inclined to use a sequential way of writing and reading; it can be better understood word-by-word, sentence-by-sentence. Long sentences can be accepted, and itemized writing is accepted as well. The Chinese are inclined to use a graphical way of writing and reading, including thinking. They understand the meaning of the sentence by combining every word around it, not just word by word sequentially (Chang 1993). The basic reason for this is as follows. There were around 5,000 characters in ancient China 3,000 years ago, and it has remained so today. We simply cannot memorize so many different picture-like characters. Five-thousand might be considered a limit for human comprehension. As a result, it is not uncommon for one character in Chinese to bear several different meanings and sounds. How it is read and realized depends on the segmentation with neighboring characters. If we do not take into account the basic difference between Chinese and English, and translate the manual directly from English, or even use the original English version, it will very likely lead to difficulties in reading comprehension, thus lowering the willingness of pilots to abide by the manual.

The difference between the reading processes of Chinese and English is also very important. The reading processes of a language can be separated into four stages: decoding, literal comprehension, inferential comprehension, and metacognition (Frederiksen 1981; Gagné et al. 1993; Just and Carpenter 1987). The apparent difference between the reading processes of Chinese and English is at the stage of decoding. From research in Chinese language processing, it can be noted that the word, instead of the character, is a complete conceptual unit; however, the word boundary in Chinese lacks visual–spatial salience for word perception during reading. There have been several studies on the effect of word boundary adjustment in Chinese reading. But the results of these studies show that the perceptual salience of Chinese words makes no significant difference for reading time and reading comprehension (Hsu 1998; Yang 1998). This can be explained by the fact that determination of word segmentation in reading Chinese is a skill that usually cannot be improved by simple adjustment.

An accepted view of memory capacity is seven units ("plus or minus two") (Miller 1956). This is because the attention span of humans is limited (Luria 1973). In psychology, attention represents an organized human mental activity possessing some degree of directivity and selectivity. Of the many stimuli that reach us, we respond to only those few which are important and correspond to our interests. In language processing, this kind of human activity is obviously at a level higher than word segmentation. On the other hand, it is also lower than the level of the sentence. After using these pictorial characters for thousands of years, Chinese people have developed a special kind of proneness in reading. It is intended here to determine the habit of chunk grouping by using the number of Chinese characters between punctuation as an index above the word segmentation level and to examine its influence on the understanding of Chinese.

One written Chinese character usually bears many different connotations. When combined with different characters, it will have a different explicit meaning. Usually, Chinese readers have to segment the characters properly to obtain words to construct sentences. Between word and sentence, Chinese people have to further group several characters or words, or their combinations, into one concealed unit in order to comprehend. In other words, readers have to group several words or items into a chunk before the sentence is formed. In the long history of China, there was neither punctuation nor sentences in ancient Chinese writing, because at that time, the characters were written on sticks of wood or bamboo. There is just no such luxury to waste space on punctuation. Hence, all the characters were written continuously without stop. Because of a limited attention span, Chinese people had to break between every several characters when reading, and this continued to be true even after the "Plain Chinese Movement" of 1920–40, when punctuation was first advocated and widely adopted. Consequently, everybody became accustomed to reading according to this hidden pattern. This underlying structure between word and sentence is called the "grouping pattern." It represents the limitation on the number of picture-like characters that the Chinese mind can process at once.

In traditional Chinese reading, a single word is usually not the basic meaning unit because its meaning is still not certain. In a sentence, one has to segment between words to confirm which word or words associate with it. This is called word segmentation, similar to phrasing in English. Once the segmentation is finished, there will be several meaning units formed. However, in order to combine these units to form a sentence, a higher-level segmentation will still be needed according to the Chinese mode of thinking. This reading pattern has existed for thousands of years. After punctuation was adopted, people then started to use punctuation marks to clarify this pattern when reading. And, it is called "grouping pattern" here.

In order to give a general description of this grouping pattern, the first step is to find statistics on the number of Chinese characters between punctuation in existing Chinese articles. Punctuation represents a pause. It also delineates the boundary between elements, or units. It is used to help construct the meaning of a sentence. The articles used consisted of 37 ancient Chinese articles from the Chou Dynasty to the Qing Dynasty (1066BC–1911AD), representing ancient Chinese writings. Also included were 16 articles written from 1920–40, at the beginning of the use of punctuation in Chinese writing. Thirty-six textbooks widely used in Taiwan elementary schools to teach Chinese, arithmetic, and science were also collected. Fourteen pages of contemporary Chinese writing from present-day newspapers were also gathered, excluding the advertisements. All of these articles were put together and the number of characters between punctuation was calculated and accumulated. Although all the articles before the "Plain Chinese Movement" in 1920 originally lacked punctuation, it was added by scholars at that time according to the traditional Chinese reading and understanding habit (Jing and Lee 2001).

The results of these statistics are given in Figure 9.6. The vertical axis represents the percentage of each number of characters, while the horizontal axis shows the number of characters. From this figure, it is clear that there is a predominant concentration at four characters in all the ancient Chinese articles. Most of the percentages are concentrated at the range from three to seven. There is almost no sentence with more than 10 characters. After the "Plain Chinese Movement," the essence of which was to write what was said, this concentration started to disperse. More characters began to appear between punctuation. The number of longer sentences began to increase. This distribution of the percentage has even flattened for modern newspapers. Although the distribution has changed, it should be emphasized that the use of this basic grouping pattern of four to ten items (seven, plus or minus three) has continued for almost 3,000 years. Even today, this four-to-ten pattern still represents the largest percentage in newspapers.

From these statistics, it can be seen that the pattern of four to ten between pauses is an underlying structure that has existed for quite a long time. It is still preserved today with minor modification. Thus, each pause with four to ten Chinese characters could be seen as a dominant pattern in Chinese writings. This custom is therefore defined as the "grouping pattern" in reading and writing Chinese. Chinese writings for all periods consistently reveal the same basic structure of pauses with about four to ten characters. These results indicate that Chinese people should presumably be using such a pattern of four to ten between each pause in reading, writing, and possibly information processing. Therefore, we can infer that Chinese pilots use the same type of grouping habit when reading flight operation manuals. If the flight operation manual could be written according to this type of grouping mode, keeping the number of words between four and ten as far as possible for both sentences and steps and procedures, it would reduce the cultural pressure faced by readers by altering or adjusting the steps and procedures to something that's more suitable for a Chinese graphical way of thinking. It could be assumed that it would actually help the readers to better comprehend the flight operation manual.

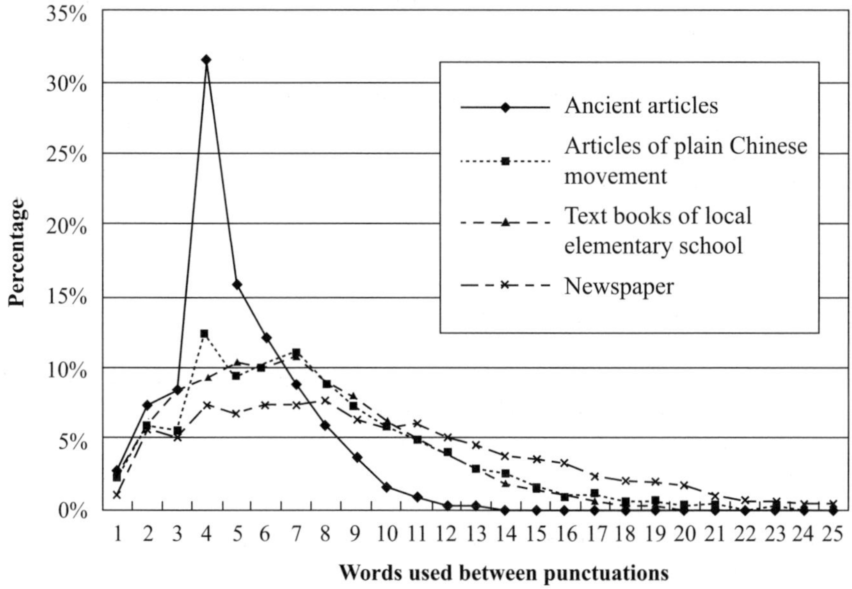

Figure 9.6 Number of words between punctuation marks in the most representative Chinese articles from the Zhou Dynasty to today

As described in Chapter 4, one of the features of Western aviation technology is procedures. In order to operate such a highly complex technological system efficiently and safely, there are many standard operation procedures established everywhere. Flight operation manuals represent one of the most typical examples. Usually, in a flight operation manual, there are many different kinds of procedures with collected items arranged in certain orders. As can be seen, "sequence" is really the key word in all of them. However, according to the Chinese reading habit, "group" is the key word for the Chinese. When a Chinese with a "group" mindset encounters a "sequential" manual, obstacles are unavoidable.

To evaluate the influence of this hidden grouping pattern on the understanding of flight operation manuals, an article from Embry-Riddle Aeronautical University about the reporting of incidents from the flight operation manual is a good example (Embry-Riddle Aeronautical University 1999). Section 2.7 "Reporting of Accidents, Incidents, Hazards, Hazardous Events" from the manual was chosen and translated into Chinese. Here we use two kinds of translation. One is direct transliteration from English to Chinese without any alteration. The other is a rewritten translation and follows the Chinese grouping pattern by adding punctuation to the sentences. It kept the original meaning, structure and grammar the same to avoid too many continuous words, and to slightly change the structure of the article by orderly grouping the listed items into several groups. In the rewritten article, the number of Chinese characters between punctuation was adjusted so as to be mostly at the range of four to ten, and the 25 listed items were grouped into five (between four and ten) categories, with each category containing not more than six (between four and ten) items to make it more suitable for the habit of readers with an ideographic mindset.

The original English article used for direct and rewritten translations are given below:

> Mandatory Reporting Events: The following list further defines those events which require a report within 24 hours.
>
> 1. When a system defect occurs which adversely affects the handling characteristics of the aircraft or renders it unfit to fly.
> 2. When uncommanded power loss occurs.
> 3. When there is fire or smoke.
> 4. When an emergency is declared.
> 5. When safety equipment or procedures are defective or inadequate.
> 6. When deficiencies occur in operating procedures or manuals.
> 7. When operating standards are degraded because of deficient operating, or maintenance, or support facilities.
> 8. When an event occurs resulting in aircraft or property damage.
> 9. When an excursion occurs, if any part of the aircraft leaves the paved surface during taxi, takeoff, or landing.
> 10. When an incursion occurs.
> 11. When an unsafe gear indication occurs or the landing gear fails to extend or retract.
> 12. Whenever aircraft landing difficulties are experienced.
> 13. When the flight crew become lost or disoriented.
> 14. When there is an exceedance of the limiting parameters for the aircraft configuration.
> 15. When communications or navigation systems fail or are impaired.
> 16. When flight diversion or return to field occurs, or when landing on wrong runway.
> 17. When a loss of braking occurs.
> 18. When the aircraft lands with less than reserve fuel remaining.
> 19. When a near-miss, ATC incident, or wake turbulence event occurs.
> 20. When significant turbulence, wind shear, or other severe weather is encountered.
> 21. When serious illness, incapacitation, injury, or death occurs to crew or passengers, if the emergency medical kit is used, or if contact is made with body fluids.
> 22. When the use of drugs or alcohol by any crewmember on duty is detected.
> 23. When a tail wind, bird/wildlife strike or foreign object damage occurs.

24. Any event where safety standards may have been compromised.
25. Any event which may provide useful information for the enhancement of aviation safety.

The rewritten article went as follows. We categorized these 25 items by putting the similar ones into five categories, and within every category there were fewer than 10 items. The five categories are: (1) mechanical problems, which includes items 1, 2, 11, 12, 15, 17; (2) human and operational problems, which includes items 9, 13, 14, 21, 22; (3) traffic and weather problems, which includes items 16, 18, 19, 20; (4) equipment problems, which includes items 5, 6, 7; and (5) miscellaneous items, which includes items 3, 4, 8, 10, 23, 24, 25. Then, we make the number of words in every sentence between punctuation in the range of four to ten. The words used are exactly the same for both the direct and rewritten translation versions. Hence, we finally changed this long mandatory reporting events list into five groups, with each group having fewer than 10 items.

After the articles were prepared, 95 sophomores and juniors from the Institute of Aeronautics and Astronautics at National Cheng Kung University were randomly chosen and divided into experimental and control groups to read either the rewritten article or the directly translated article. Subjects were selected so that they were similar in background and ability level. Their grades in Chinese on the college entrance examination were taken as a comparing reference. Subjects were instructed to read either the rewritten article or the directly translated article in 10 minutes and were not allowed to talk to each other. After they finished reading, the articles were taken away and the reading test was immediately given to determine the influence of grouping pattern on reading. Both groups were asked to answer the same questionnaire. The reading test consisted of 15 questions with no time limit. All students were given the same questionnaire and told to choose the most appropriate answers according to the content of the article they had read.

Some students did not have grades from the entrance examination, so they were excluded from the study. For this reason, only 56 subjects were valid, with 26 reading the rewritten article and 30 reading the directly translated article. As shown in Table 9.1, the average scores in Chinese on the college entrance examination were roughly the same. However, the results of the reading test showed a difference. The average grade of students reading the rewritten article was 75.13, which was higher than the average grade of the other group, 70.22. The perfect score of this reading test was 100. After being normalized with the score in Chinese of the entrance exam, the results of the reading test revealed that the reading comprehension scores for students who read the rewritten article were approximately 10% higher, as revealed in Table 9.1. Also, the t-test result shows that the difference was significant statistically. These results confirm that the grouping pattern has an effect on the reading and understanding of the flight operation manual.

Grouping pattern is a hidden structure between word segmentation and sentences. Word segmentation involves the rearrangement of only a few characters to obtain the proper word and its meaning, say, two or three. On the other hand, a sentence usually contains more than 10 characters. Especially in English, a sentence with more than 20 words is not uncommon. When it is directly translated into Chinese, it simply does not fit into the traditional Chinese reading habit. However, the grouping pattern had no influence on word segmentation. It could not affect the extraction of the meaning of words. On the other hand, it did help to integrate the meanings of words to find the idea transmitted by the whole sentence. It is obvious that if there is a step between word segmentation and sentence to help, the construction of the meaning of a sentence will be easier. In other words, it makes the route directed from word to the idea of the sentence closer. It also saves the reader time in finding the idea of a sentence. So it is suggested that fitting the article with the grouping pattern of Chinese will be helpful for readers' comprehension.

Table 9.1 Difference in reading comprehension when using Chinese grouping pattern

Results / Group	Number of students	Chinese score in entrance examination	Reading test score	Normalization with score in entrance examination (%)
Rewrite translation group	26	67.53	75.13	111
Direct translation group	30	68.67	70.22	102

The difference between reading comprehension test results is about 9%, and the difference between the accident rates of Asian and Western pilots because of a lack of adherence to procedures is similar at 12%. It might just be coincidental that this percentage difference corresponds with the difference between Asian and Western pilots in adherence to procedures, yet the similarity invites further attention. However, both the phenomena these two numbers showed are relevant to the standard operating procedure. One is about the comprehension of the standard operating procedure; the other is about adhering to the standard operating procedure. Since they are both relevant to the standard operating procedure, it's very unlikely that the similarity of these two numbers is just coincidence. Difficulties with comprehension undoubtedly affect the ability and willingness to adhere to the SOP. These two almost identical numbers actually show that when an ideographic-thinking pilot faces a flight environment based on linear thinking, the idea that potential threat originates from incompatibility simply cannot be disregarded or misunderstood. This difference does not come from Chinese pilots being less inclined to adhere to procedures. Rather, it comes from the cultural pressure for Chinese pilots to actually comprehend and abide by them when facing linear standard operating procedures from the West.

In just one single article's reading comprehension, the performance of ideographic-thinking people declined when entering into an environment built by linear-thinking people. In the global community of aviation safety, a 10% reduction in accidents is quite substantial. Besides, the difference between modes of thinking of Chinese and Westerners would not only affect how they comprehend manuals. Other things such as the organization, management, surveillance, and monitoring and operation of the entire system, and even the perception, judgment, and decision-making of flight crews would be even more affected by the difference in thinking modes. Therefore, when Chinese pilots have no choice but to face the linear surrounding that is designed and built by Westerners, the Chinese will have to face it in a very cautious and careful way. This is also true for the Western world, since this is a problem of incompatibility.

Chapter 10
Flight Safety Margin Theory

Based on the inadequacy of the flight safety theories originating from Western linear thinking, and the difficulties of ideographic thinking operating in the environment created by Western thinking, it is necessary to consider establishing a theory of flight safety using Chinese ideographic thinking from the very beginning. However, do we really need a different type of theory when we already have several good ones, such as domino, accident chain, and defense-in-depth? The answer is yes, even if purely for academic interest.

If the present theories could supply the whole picture for us to understand the entire story, or their explanations were sufficiently culture-free so that we could directly use them to understand events, solve problems and promote flight safety, of course it would not be necessary to build a different theory, just like we do not need to create a Chinese version of Newtonian mechanics. If the aviation environment in Taiwan is almost the same as that in the West, we still do not need a new type of theory since the same environment represents the same problems encountered. Obviously, the local Taiwanese environment is quite different from the West. The whole system of aviation, including theory, design, manufacture, production, operation, maintenance, training, management, control, regulation, and even investigation are all created by Westerners. They are suppliers as well as users. However, Taiwanese airlines are users only. In many ways the entire system, including its underlying logic and tacit assumptions, is a "black box." That means the Taiwanese operators have to face a huge "black" system. Although, all they need to do is simply make themselves familiar with the system. However, in this case, cultural distance will determine how difficult this will be.

Finally, if we perceive flight safety from the same angle as Western thought, it would be unnecessary to think about creating a different theory. However, even if we want to view flight safety in the same way as the West, it is still very difficult for us to achieve. Originating in an alphabetical system, the Western linear mode of thinking stresses sequence and values logic, with analytical capability as its specialty. On the other hand, the Chinese ideographic mode of thinking, originating from Chinese writing characters, is holistic, stressing equilibrium between components. Its special capability – different from its Western counterpart – is insight, i.e., finding system features given only very little information. These two modes of thinking are not just different; they are in different dimensions. They are orthogonal. Moreover, this fundamental difference has existed for at least 2,000 years because of geographical environments. As long as Chinese people still use their writing characters, and Western cultures continue to use their alphabets, the difference will persist in the foreseeable future. As a result, we will never be able to perceive flight safety from the same perspective as Westerners, just like the cat can never see the world from the perspective of the dog. Consequently, it would be quite helpful to have a theory derived from the Chinese ideographic mode of thinking, i.e., to see our flight safety with our own eyes, to guide us about the planning, design, operation, control, and management of the local adaptive aviation system to pursue the efficient allocation of resources and safe operation of each party. If the theories from linear thinking are not in accordance with our needs, how can we create a theory for flight safety from the Chinese ideographic mode of thinking to meet our needs?

The flight safety margin theory proposed here represents a preliminary, even primitive, attempt to explore the possibility of creating a non-Western scientific theory. We can see it is at least feasible, considering the existence of Chinese medicine, a surely non-Western medical theory that is beginning to be accepted in parts of the West. But this theory is also far from achieving any kind of accomplishment. It is only a preliminary trial.

Rationally, we will need three steps to do this:

1. Based on the holistic Chinese ideographic thinking, search for the appropriate perspective for flight safety.
2. In view of the graphical nature of Chinese characters, try to establish a two-dimensional representation of flight safety.
3. After abstraction of the two-dimensional representation, geometrize flight safety.

Basically, the appropriate perspective for flight safety really means using our own eyes to see our flight safety since we have to solve our own problems. From the global data for aviation accidents, it is clear that flight crews are at the center of flight safety. The crews are the persons who fly the planes and are also the last link of the accident chain by Western thinking. Once the airplane leaves the ground, it is almost the case that only the crew can determine the fate of the plane. Generally speaking, according to investigation reports, in many accidents, even before the airplane left the ground, the errors were already there and kept on evolving. In these cases, if the crew had enough information and knowledge, to make correct judgments and take actions promptly without delay, most of the accidents would have had more of a chance of being prevented. In other words, the pilots were indeed crucial to the flight safety. It is even more so in Taiwan.

In Taiwan, the facts for aviation are as follows: the airplanes are designed and manufactured in the West. The standard operating procedures are established by Westerners. Maintenance regulations are also written from a Western perspective. Our pilots are almost all trained by Western education. Even accident investigations follow Western standards. Obviously, in the global aviation system, we are only an operator. Hence, based on the limited resources in Taiwan, we would have to begin with defining what the flight safety is to us because it would be quite unreasonable if we use the same definition of flight safety as the Western perspective.

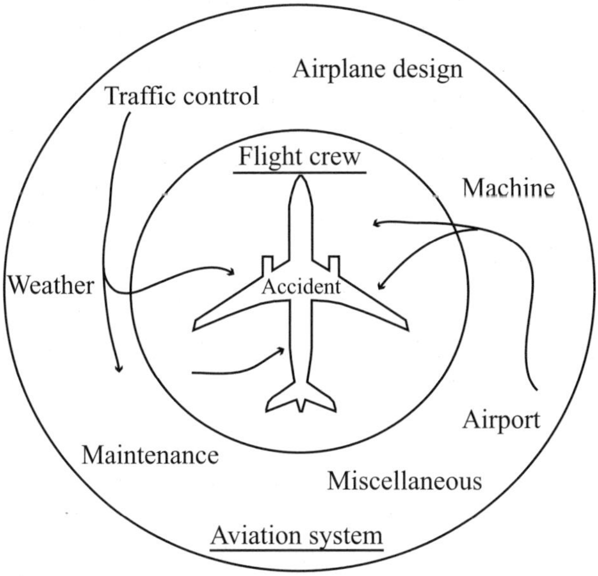

Figure 10.1 The concept of flight safety from the ideographic mode of thinking perspective (compare with Rasmussen 1997: 190)

Since we are only the users, flight safety would simply mean "the pilots safely operate the airplanes." The basic philosophy of the government aviation policy in Taiwan would thus be: use all the available resources to help pilots safely operate airplanes, and the corresponding purchasing policy, maintenance, operation, management, control, and monitoring, together with the crew selection, training and licensing, can all be toward the same goal. Consequently, any local flight safety theory would have to put pilots at the center.

From a Taiwanese perspective for flight safety, considering the complex nature of the aviation system, there are multiple events occurring at every moment. A two-dimensional theory visualizing flight safety as a picture instead of a line would be much more suitable to highlight the feature of ideographic thinking from Chinese characters, regardless of the sequential-type theories. Figure 10.1 portrays the graphical theory with pilots at the center. The outer circle represents the whole aviation system. All the factors which can potentially affect flight safety are included, such as design, manufacture, maintenance, operation, training, monitoring, together with air traffic control, airport facilities and management, and meteorological conditions, as well as corporate culture, social conditions, economic and political climate, etc.

The inner circle represents flight crew. All the factors related to pilots flying are included. They cover factors such as personality, capability, skill, knowledge, experience, attitude, cognition, mindset, decision-making, and behavior. It should be noted that all these factors are regional, i.e., pilots from different regions have different characteristics, and will be influenced by the local environment. The airplane at the center represents the aircraft accident. The airplane is enclosed by the circle of the pilot, representing a potential buffer provided by pilots between the accident and the aviation environment faced by pilots, since pilots can provide the final protection to the aircraft. If pilots handle the situation correctly, the accident can be stopped. Otherwise, the event can penetrate the protection of the pilot and reach the accident.

Aircraft accidents are seldom caused by a single factor. Most of the time, there are multiple factors evolving simultaneously. They are often interrelated and sometimes mutually reinforcing to alter the path of advancement of the accident nonlinearly. Finally, one factor among them reaches the threshold and results in the accident. All of those factors directly and sequentially related to it become the primary causal factors of the accident. All of the factors linked by any evolution process of an abnormal situation can be represented by an "event curve." The event curve with the final factor as the accident is called the accident chain. The event curve can start from any point in the whole system as well as the pilots, representing the first domino or link. The flight situation was then changed and started to evolve. Graphically, the event curve begins to march and will certainly encounter the circle of the pilot. At this moment, the pilot faces an abnormal situation. If the pilot handles the situation correctly, this event will be resolved and the accident prevented. The curve is stopped without reaching the airplane. If the pilot made mistakes, the event curve will penetrate the defense of pilots and reach the aircraft, representing the accident, or if the pilots were not capable enough to stop the advancement of the event curve, the accident will also take place. Of course there could be several event curves evolving simultaneously and encountered by the pilots.

If the pilots are not well trained, the circle of the pilot will shrink and move closer to the airplane, representing a situation where the pilots cannot provide sufficient protection against events – because wherever the event curve encounters the pilot, it is already very close to the accident occurring. In this case, the buffer from the pilot is very thin. On the other hand, if the pilots are very capable, the corresponding circle will be large enough. It represents that after any event takes place, the pilot will have plenty of room to protect the airplane to stop the curve from marching into the airplane. The possibility of an accident is then lowered, just as the airplane is protected by a large buffer zone. However, if the flight environment has deteriorated as a result of factors from the aviation system, e.g., interference from politics such as setting up an airport in a place which is not suitable, opening a route which is not allowed, or improper pressure from media, as well as other factors in the company itself, such as an unhealthy corporate culture, the company does not value safety properly, inadequate maintenance equipment, and other matters, the pilot will face a dangerous environment created from the whole system. All of these factors will compress the space in which the pilot can safely operate the airplane. As can be seen in Figure 10.1, it resembles the size of the outer circle. If the outer circle is reduced, wherever the event curve starts, it will quickly meet the inner circle representing the pilot, and compress the space for the pilot to safely fly the plane.

Hence, from the size of the inner circle representing the capability of the pilot, together with the size of the outer circle symbolizing the health of the flight environment, a new concept for flight safety can be extracted. Figure 10.1 strongly suggests a concept of "space" to represent flight safety. This safety "space" illustrates the "room for the pilot to safely operate the airplane." It also represents the difficulty of the event curve reaching the airplane, i.e., an accident. This we characterize as the flight safety margin (Jing 1998).

From the perspective of flight safety margin theory, we can clearly understand that to promote flight safety really means to enlarge the margin of safety, i.e., to enlarge the space for the pilot to safely fly the plane. As for how to enlarge the margin, there are two ways of doing this. First, improve the capability of the pilot. This will enlarge the inner circle in the Figure 10.1. From the viewpoint of improving the capability of the pilot, it includes selection, inspection, training, monitoring, etc. Everything which can promote the overall capability of the pilot and prevent the event curve from advancing are included in this category. Second, reduce the pressure from the aviation system. Anything that can enlarge the outer circle will be included in this category. The flight environment for the pilot should always be kept friendly to the pilot. For example, using mobile phones in the airplane, high-rise buildings around the airport, birds flying close to the airfield, lack of navigation aids, social pressure, and interruption from politics, among other matters, will all compress the margin of safety for the flight operation of the pilot and should be avoided as far as possible. In addition, other factors from the airline as well as the aviation authority will also effectively influence the room for the pilot to safely operate the airplane.

Despite these local factors in Taiwan, the others from outside can still possibly compress the margin of safety. These include the basic assumptions in the design of the airplane, automation, standard operating procedures, and even training such as CRM or regulations, and other matters. The foreign manufacturers of the airplanes could not possibly understand the mindset of the Taiwanese pilots. The products, whether hardware or software, based on their own assumptions, will not fit generally into the behavioral pattern of the local pilots. In case of emergency, how the Chinese pilots will react is often outside the mindset of the foreign manufacturer. Therefore, these discrepancies frequently create a mismatch between the machine and pilots, and produce events and further foster the evolution of the events. All of these factors implanted into the aviation system will compress the safety margin of the pilot's operation for flying.

The flight safety margin is a two-dimensional, instead of sequential, theory with the pilot at the center, from the user's point of view, considering the complex nature of aviation to delineate the occurrence and evolution of the events. Also, this theory can provide the means to promote the capability of pilots together with improving the flight environment to lower the pressure on the pilots and enlarge the margin of safety, i.e., provide room for the pilot to safely fly the plane. If this can be done, it will be the best policy for accident prevention, the ultimate goal for flight safety. Consequently, the flight safety margin theory can provide a holistic picture for flight safety, so as to guide us about many aspects for improving flight safety.

Having a graphical representation of flight safety from Chinese ideographical thinking, the third step is to geometrize the flight safety margin through abstraction. Compared with the alphabetical system, the Chinese characters are basically geometrical instead of analytical. Having been nurtured by the graphical characters for such a long time, Chinese minds are more effective dealing with abstraction of figures as it can be seen from the classical Chinese painting. Also, there have been only two internationally famous Chinese mathematicians in the last hundred years. They are Shiing Shen Chern and Shing-Tung Yau. Yau is also the only Chinese winner of the Fields Medal, the Nobel Prize for mathematics. It is quite surprising that both of their contributions are all in the field of geometry. However, it is not purely coincidental, because it is very logical. Geometry is closer to Chinese ideographic thinking compared with other fields of mathematics. Hence, to create a flight safety theory with Chinese thinking, geometrization is the most natural approach.

To create an academic theory, i.e., one that can be taught and applied in multiple contexts, abstraction is a must. It is always necessary to extract the general rules out of various complicated phenomena, then to explore the operations of the rules and create a descriptive framework for them. The process is the same for both the linear and ideographic modes of thinking. The difference lies in the dimension in which the abstraction is executed. In Western thought, the abstraction is performed along a line. That is the reason why those ancient Greek philosophers came up with the concepts such as ideas, the basic constructing element of thinking along a line (Jing 2012). However, for the Chinese, the abstraction is executed two-dimensionally. There is no concept such as "element" in Chinese culture. After abstraction on a plane, the set of information remaining is called a characteristic or feature, an interrelating group of information in two dimensions. Of course, there could be many features in a picture.

A simple example of abstraction in one- and two-dimensional thinking is as follows. As everybody knows, there are four basic constructing elements of the universe in ancient Greek philosophy. They are water, air, fire, and earth. On the other hand, as understood generally, the Chinese five "elements" are water, fire, earth, metal, and wood. Quite surprisingly, three are the same. This is the reason why many scholars believe ancient Chinese science had already reached the same level as ancient Greece. However, there also exists debate about the English translation of the five Chinese "elements" because the meanings of these "elements" are obviously far exceeding the English equivalent for "element." Of course, the so-called five Chinese "elements" are actually not elements but features or characteristics, because in Chinese thinking those five "elements" can be used in many different areas, from medicine to climate as well as society.

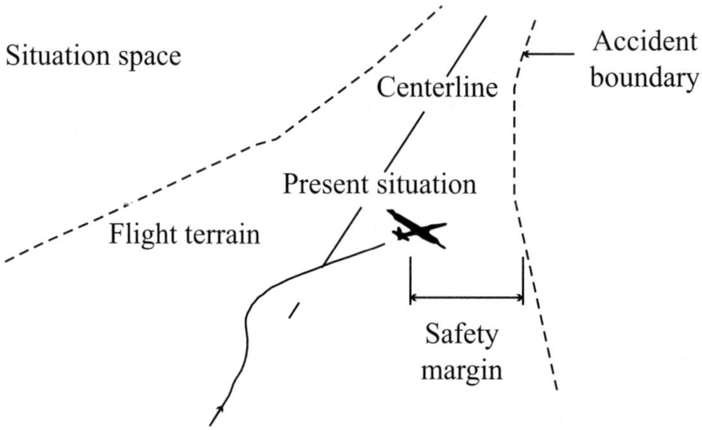

Figure 10.2 Abstract schematic diagram for flight safety margin theory

If anyone comprehends the Chinese five features as Western elements, they would definitely be confused by why these elements can be used to explain the operation of the human body as appears in Chinese medicine. Hence, water, fire, earth, metal, and wood are features, not elements. Any phenomenon with the same features can be called under that name, no matter where the phenomenon takes place. Consequently, the Greek water, air, fire, and earth are units, while the Chinese water, fire, earth, metal, and wood are features or aspects, and they are entirely different. Therefore, the difference between the linear and ideographic modes of thinking is quite obvious and easy to comprehend.

It is quite clear that the abstraction in two dimensions really means looking for a feature, a simplified set of information, from the whole field of the object, with emphasis on correlation, which can grasp the main essence of the whole story in a picture. It is also discovered in the study in Gestalt psychology (Arnheim 1969). Hence, the final step for flight safety margin theory will be to look for the geometrical feature from Figure 10.1. The key concept to portray safety from Figure 10.1 is the space or room in which the pilot can operate the airplane safely. That is to say to define a "space" in which the pilot operates the airplane is the first step. Of course, it will not be the physical space. Since situational awareness is very important to pilots, the space is defined as "situation space." Situation space is an imaginary space which should include all the factors affecting flight safety, whether it is from the aviation system or the pilots. In the development of flight safety margin theory, an abstract situation space is then defined first, as shown in Figure 10.2. All the factors related to flight safety constitute the space representing the set of all possible situations. Thus, take the SHELL model (**S**oftware, **H**ardware, **E**nvironment, **L**iveware (social environment), and **L**iveware (operators)) (Hawkins 1987) as an example; the situation space includes all the parameters, from software and hardware to environment, as well as human factors. Of course, since the situation space is an abstract space for depicting flight safety, the dimensions can be chosen as necessary, as long as they can reveal the features of flight safety: the dimensions of a feature space are always lower than the original space. Moreover, in this space, any combination of parameters is represented by a point. Any point in this space represents a specific situation of flight. Since there exist constraints from physical laws about any real flight, realistic flight situations constitute a hypothetical virtual terrain in this space, which is called the flight terrain, a subspace with fewer dimensions.

The other reason for defining situation space is as follows. The core concern of flight safety is the human error in flight operation. Therefore, it is appropriate to establish a theory of flight safety from the point of view of operators. Operationally, safety can, in principle, be interpreted as room for error. Flight safety margin can thus be delineated as the room for error or mistake in flight. However, the same error occurring in different situations will induce different degrees of danger. Take a simple example: if the conflict between human and aircraft, as in the case of Nagoya, took place at an altitude of 10,000 feet instead of 1,000 feet, the outcome would have been totally different since the crew would have had much more room to recover the aircraft from potential accident. In other words, finding out how dangerous the error or mistake is, without referring to the situation, is extremely difficult if not impossible. Consequently, the situation has to be defined first before the discussion can proceed.

For any real flight, it can be considered as the continuously changing situation. In other words, it is the time-varying combination of different parameters. Hence, in the situation space, every flight can be regarded as a time-varying continuous curve traversing the virtual flight terrain. For every flight mission, there is a set of perfect, standardized situations required by the related regulations. If all the situation parameters work perfectly, the corresponding curve should represent the safest flight, which is defined as the centerline on the terrain in the situation space. Any normal flight is thus a continuous curve wandering around the centerline because the crew will do their best to maintain the aircraft as safely as possible, i.e., adjusting the situations, combinations of flight parameters, to avoid deviating away from the centerline, standardized situations required by regulations.

There are also certain combinations of the parameters which constitute the situations of an accident. They are represented as the accident boundary. If the situation deviates away from the centerline for whatever reason (weather, machine, maintenance, operation, or other matters), it means the risk of an accident is increasing. Naturally, the curve will be closer to the accident boundary. Thereafter, if the situation has reached the accident boundary, it means the situation is already outside of human capability to recover. The accident becomes inevitable. Hence, the accident boundary really delineates the limit of flight situation. Of course, the accident boundary will vary according to the different phase of the flight. When the airplane is closer to the ground, the accident boundary will also be closer to the centerline. It means that it will be much more likely that the accident will take place. Having the centerline and accident boundary on the flight terrain in the situation space, every possible flight would then be a curve advancing between the centerline and the accident boundary, as shown in Figure 10.2.

Through abstraction, the holistic picture of flight safety in Figure 10.1 is now drastically simplified in Figure 10.2. The concept of using "space" to represent safety in Figure 10.1 can also be greatly simplified as a "distance." In Figure 10.2, for any instance of any flight, the flight situation is a point on the curve which can be called the "present situation." From this point, there is always a distance to the centerline as well as to the accident boundary. The distance between the present situation and the centerline represents how close the flight is to the standardized situation. On the other hand, the distance to the accident boundary delineates how close the present situation is to the accident. At any given instance, this "distance" can thus be used as a safety measure because it implies how far the present situation is from the occurrence of an accident, which clearly means a margin of safety. Hence, the distance is thus defined as the "flight safety margin," as shown in Figure 10.2. Once the flight safety margin is defined, the imaginary concept of safety space or safety room for the pilot to operate the airplane is then simplified as a distance. According to the definition, the larger the margin, the safer the situation because that means the pilot has more room to operate the airplane safely. If the margin is very small, it represents the situation is already very close to the accident, i.e., not much room left before an accident occurs.

If the flight safety margin is defined in terms of distance, it would then be possible to find a way to measure it. The first difficulty for finding a measure is that it is almost impossible to figure out the combination of flight parameters representing the accident boundary without sufficient data. The combination of data points is also believed to be unlimited and impossible to collect. The second difficulty is that safety is a perception problem and will vary from person to person. And, it is also believed that is impossible to establish an objective and precise mathematical definition for safety at the present stage. This is thought to be the fundamental reason why it is so difficult to study flight safety quantitatively. However, the way to overcome these difficulties could possibly be very simple. If it is almost impossible to "measure" the distance between the present situation and the accident boundary, we can "measure" the distance between the present situation and the centerline, which can be defined almost exactly since it is required by standard operating procedures. At least, if not exact, the measure is definitely feasible. For the second difficulty, it can be solved by expert systems (Giarratano and Riley 1994).

Expert systems are a branch of artificial intelligence (AI) that make extensive use of specialized knowledge to solve problems at the level of a human expert. In other words, it is an intelligent computer program that uses knowledge and inference procedures to solve problems that are difficult enough to require significant human expertise for their solution. To have such a system, the engineer must have a dialogue with the human expert having knowledge or special skill in a certain area, which is not known or available to most people, to extract explicit knowledge from the expert and code it into the computer system to emulate the cognitive and decision-making ability of the human expert. The process is also called knowledge engineering. In general, the expert system is composed of two parts: knowledge base and inference engine. The knowledge base contains the knowledge with which the inference engine draws conclusions. In this study, a neural network is used to build the corresponding expert system.

A neural network (Haykin 1999) is basically a mathematical method with a learning capability. Knowledge from experts is coded in an appropriate form for the computer to learn, also a process of knowledge engineering. After the knowledge is transformed into the proper representation, e.g., training examples, it can be learned by the neural network through adjusting the synaptic weights connecting neurons in the network. The training examples for the computer to learn should incorporate the unknown causal relation of the relevant phenomenon given in the form of input–output. Thereafter, once the neural network learns the knowledge, the knowledge is not presented through a precise mathematical formulation as usual but stored in the connections of the network. Moreover, the causal relation in the neural network is not limited to the ordinary "if … then" type formulation as given in many scientific laws based on formal logic. Also, the causal relation can be of multiple inputs and multiple outputs. Hence, the neural network can learn many different types of knowledge, either in sequential form or in network form, not to say parallel form. Compared with traditional scientific theories, the neural network is closer to the operation of the human brain in the way the knowledge is formed.

Surprisingly, the basic philosophy of neural networks is much closer to the Chinese ideographic mode of thinking than the Western linear mode. This is not purely by chance. Linear thinking, ideographic thinking, and even neural networks are all explanatory frameworks to portray the operation pattern of the human brain. The key feature of a neural network is the representation of knowledge by synaptic weights connecting neurons instead of neurons themselves. The neural network is thus closer to ideographic thinking because the inference procedure of the neural network is of the networking type instead of the sequential type, with emphasis placed on the connections. This is the most fundamental difference between linear and ideographic thinking, since linear thinking focuses on the substance or entity, with the thinker excluded, as described by Aristotle, while ideographic thinking focuses on the correlation among entities, frequently including the thinker.

Even with expert systems, it is still hard to perceive the "distance" between the present flight situation and the centerline. Some elaboration has to be made. Since safety represents, in principle, the room for error, the "distance" to the accident boundary can be perceived as the needed effort or performance to neutralize the error and recover the flight back to the centerline. During flight, if some error took place, the flight situation would move closer to the accident boundary. At this moment, any corrective measure or defensive strategy will try to pull the situation away from the accident boundary and back to the centerline. If the defense can be accomplished easily, it can be said that the present situation does not deviate too far from the centerline and the distance to the accident boundary is still very large. On the contrary, if the situation is already very close to the accident boundary, the performance needed to recover the situation back to normal operation will definitely not be minor. Moreover, if the errors have already accumulated to a certain stage which is beyond any possibility for human recovery, the accident is thus unavoidable. The corresponding situation then locates right on the accident boundary. Concluding the above discussions, the needed performance to recover the present situation back to the most standardized situation could be a feasible concept to delineate the "distance" between the present situation and the accident boundary geometrically, and can thus be used to define the flight safety margin.

With a neural network, the restrictions originating from individually different perceptions about safety can be overcome because the neural network can be considered as the tool used to learn the knowledge, experience and also perception of human experts. In principle, the unknown causal relation between situation and safety, no matter from whose perception, has to be incorporated in the training example and expressed as the input–output form for the computer to learn. Naturally, a set of flight situation parameters represents an input, or, more specifically, a time-varying vector of input. The output is designed to be the safety margin of flight. Of course, given a set of parameters, e.g., take it as simple as possible, height, speed, angle of attack, descent rate, flap setting, and others, how safe this situation appears will differ from person to person. As long as the different perceptions can be presented in the input–output form, it can always be learned by the neural network. If this can be done, the corresponding expert system can thus supply the unknown causal relation, fitted from the average of the perceptions from many experts, between the situation and safety margin by using a neural network. The entire logical foundation of the flight safety margin theory with expert system can then be completed as shown in Figure 10.3 (below). As shown in this figure, given any kind of flight situation including threat (Helmreich et al. 1999), through the established neural network incorporated with the learned causal relation from experts, the flight safety margin can be calculated as the output.

To establish the corresponding expert system, a questionnaire was needed to collect the expertise of the pilots to supply their knowledge about safety. As explained previously, the present study utilizes the needed performance to recover the flight situation back to the most standardized one to represent the distance between the present flight situation and the accident boundary. Only the pilots understand how much effort they have to devote under the given situation. This questionnaire was designed for the pilots to reasonably classify their perception about the needed performance. After interviewing some senior captains, the needed performance was categorized into two main parts: physiology and psychology. The physiological portion was further partitioned into agility and skill. The psychological portion was partitioned into experience and knowledge. In total, there are four sub-categories in the questionnaire. These four sub-categories were chosen by adding the basic instinctive response in addition to Rasmussen's skill–rule–knowledge classification of human performance (Rasmussen and Jensen 1974).

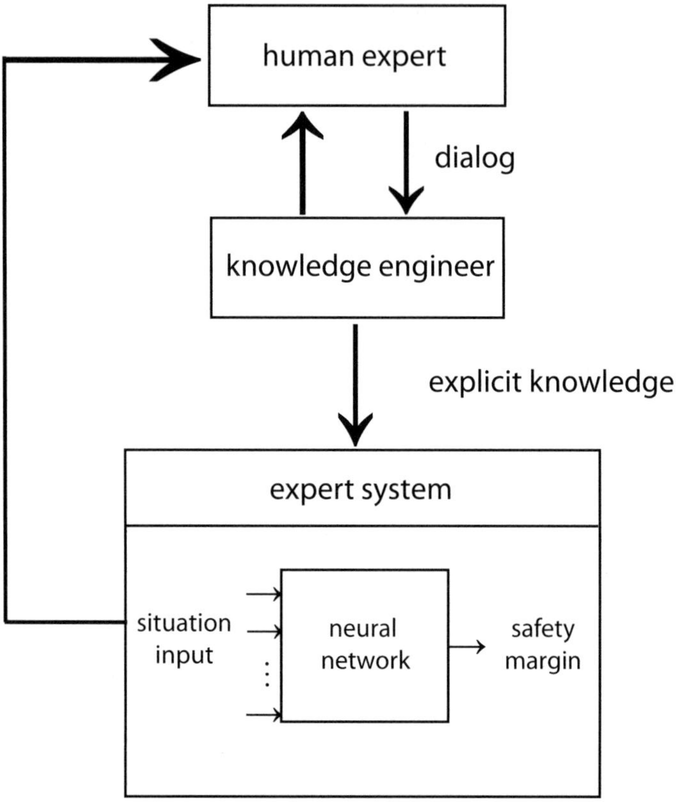

Figure 10.3 The expert system of Flight Safety Margin Theory using neural network

There were also four items in each sub-category to reveal the magnitude of the needed performance. They were low, medium, high, and beyond. "Beyond," in this questionnaire, represents that the situation has moved beyond human capability to recover, i.e., in this case, the given situation is already out of human control (Sheng 2006).

The objective of the designed questionnaire is to collect the perceptions of the experts about the severity of any given flight situation. Once this is done, the perceived severity has to be transformed into the safety margin. The more severe the perception is, the closer one moves to the accident boundary. The flight safety margin thus bears an opposite meaning to the severity. Hence, the flight safety margin can be defined as the opposite of the needed performance for recovery, i.e., the perceived severity. In order to be understood easily, the flight safety margin duplicates the definition of probability to be within zero to one, as follows:

$$\text{flight safety margin} = \frac{1}{\text{needed performance for recovery}}$$

$$= 0 \sim 1$$

0: zero safety margin, accident occurred

1: 100% safety

To obtain a number between zero and one, the needed performance for recovery among each sub-category and its magnitude has to be adjusted through the learning process of the corresponding neural network. The zero needed performance is set at 1 and the largest is set at infinity, which is represented as "beyond" in the questionnaire. By using the designed questionnaire and the definition of flight safety margin, the training examples can be prepared with the given flight situations as inputs and the safety margin as outputs, a number between zero and one representing the percentage of safety. By collecting enough examples from real flight data, the training process can be accomplished without any difficulty. By completing this, for any given flight situation, the neural network will have the capability to supply a quantified safety measure representing the averaged perceptions of experts.

For the neural network to learn reasonably, the training examples have to cover a large enough scope regarding safety and the corresponding flight situations. The real normal flight data as well as the most dangerous data have to be in the pool of training examples. One typical safe flight was chosen where 10 instances were selected for the preparation of examples of an airplane descending from 3,000 feet to landing. On the other hand, for the neural network to have the capability to supply a safety measure regarding the accident boundary, it would be better to use flight data from a real accident. The data from the Da-Yuang case were used to provide the corresponding information. Ten representative instances in the final approach, from 3,000 feet to the occurrence of the accident, were picked for use in training. Because of the limitation of the flight data, only 17 situation parameters from the flight recorder were used. They are speed, altitude, pitch angle, vertical g, angle of attack, engine EPR1, engine EPR2, engine RPM1, engine RPM2, bank angle, descent rate, distance to airport, glide slope (dot), localizer (dot), configuration, gear, and autopilot. In total, 17 parameters in 20 instances together with the corresponding questionnaire constitute the material for the survey. The local pilots asked to complete the survey all have experience in flying the same type of aircraft as in the Da-Yuang case. Unfortunately, there are not many such experts, only 22, for us to interview. In total, there are 440 training examples for the neural network to learn. As for the construction of the neural network, a multilayer network is used. To incorporate the knowledge from the four sub-categories in the questionnaire, the network used is 17-9-4-1. The learning scheme used in this study is error correction learning with back propagation.

Case Studies

Because of the difficulty in obtaining data from real flights, only four different cases were used in this study to validate the proposed theory. They are normal flights, landing in heavy fog, the Nagoya accident, and the Da-Yuang accident.

1. Normal Flight

Normal flights represent flights without any FOQA event. None of the situation parameters were over the limit set by the management. There are in total 39 sets of normal flight data in this study. The corresponding safety margins varying with time were plotted in Figure 10.4. In this figure, the changing safety margins for the final approach were shown. In total, 39 curves were given in this figure. As can be seen, starting from 3,000 feet above the ground, the situation is quite safe as perceived by the local pilots. The safety margin is around 0.98. As the airplane gradually descends, the margin of safety is compressed accordingly. The process is very smooth. When the airplane is closer to the ground, the corresponding safety margin is lowered more drastically, similar to parabolas – the closer the airplane is to the ground, the quicker the safety margin is compressed. That represents, in the minds of the local pilots, that the variation of the room for error is not linear with respect to the altitude. At touchdown, the average safety margin of 39 flights is 0.531. Clearly, it cannot be zero since zero illustrates the flight situation being already on the accident boundary. Also, the variation of safety margin becomes greater when the airplane is closer to the ground. It delineates that safety is becoming more and more uncertain and sensitive.

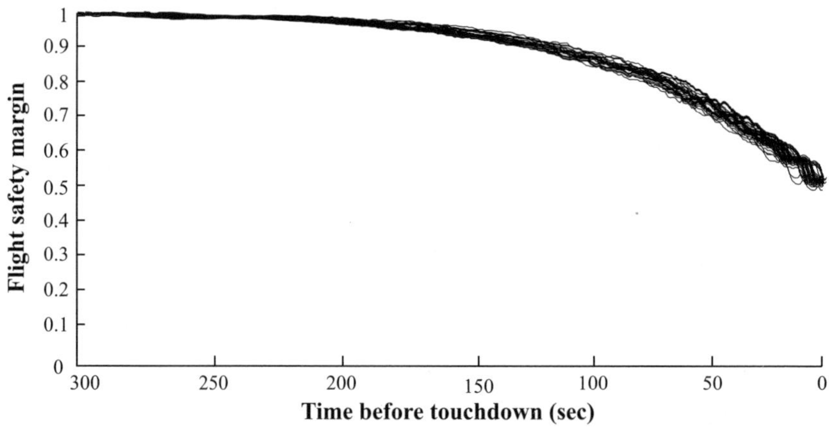

Figure 10.4 Variations of flight safety margins of 39 normal flights

A minor change in situation parameters will result in a fairly considerable alteration of the safety margin. It also represents that any minute error or even carelessness in the maneuver of pilots when the airplane is close to the ground will more likely produce an unexpected outcome.

2. Landing in Heavy Fog

One FOQA data set from a safe landing in heavy fog was chosen here to test the validity of the proposed theory of flight safety margin in a more obscure case. According to the regulations, the visibility permitted landing in this case. Again, data starting from 3,000 feet above the ground to touchdown were the inputs of the present expert system. After calculations, the changing of the corresponding safety margin was also given in Figure 10.5 with the average of 39 normal flights as a reference. As can be seen, the difference of the safety margin compared with the normal flights is quite obvious, although the general tendency is still almost the same as the previous cases, close to one at 3,000 feet, it gradually becomes lower as a parabola when close to the ground. At touchdown, the calculated flight safety margin is 0.483, 9% lower than the average of normal flights. At 3,000 feet, the influence of the heavy fog on the safety margin was not significant. At around 1,000 feet, the safety margin started to deviate from that of normal flight, and the deviation was gradually enlarged, representing the influence becoming more and more severe. A 9% difference at touchdown is relatively large, indicating that the influence of heavy fog on safety is something which cannot be overlooked.

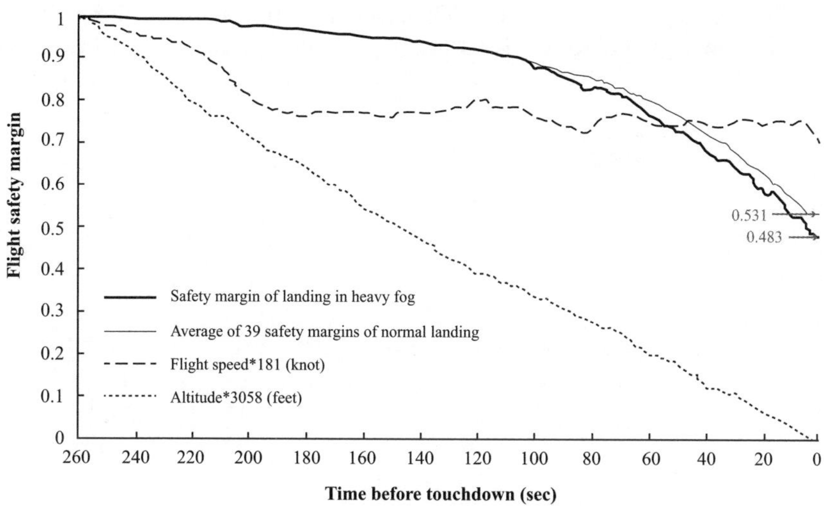

Figure 10.5 The flight safety margin of a flight landing in heavy fog

The changing of the flight situation parameters were checked to reveal the hidden cause of the difference in safety margin. It was found that, in general, the altitude of landing in heavy fog is a little higher than the altitude of a normal landing. In addition, the corresponding flight speed is a bit lower than that of normal flight. Also, the airplane speed at touchdown in the present case was lower than the speed of normal landing. The reason for the lower speed is from the thrust of the engines. After checking the engine pressure ratio, it was found that because of the heavy fog, the engines were operated at a slightly lower thrust compared with normal flights. To maintain adequate lift at lower speed, the angle of attack has to be increased to compensate for the loss in speed. After checking the data of angle of attack during landing, it was found that the angle of attack did have a higher value – greater than 10 degrees. The corresponding data for a normal flight were all less than 10 degrees for most of the time during landing. According to these situation parameters, considering the higher altitude, lower speed, lower thrust, and higher angle of attack, it is almost a certainty that during landing in heavy fog, the crew flew the airplane more carefully and conservatively to compensate for the loss in visibility due to the environment. This is the reason for the lower safety margin in the case of landing in heavy fog.

3. The Nagoya Accident

To validate the present theory having the capability to reveal the variation of safety in a real accident, the Nagoya case was chosen to be the demonstrating example. During the training, the corresponding data from the Da-Yuang accident were compiled in the training examples. The resulting expert system will be able to extract safety cognition of the local pilots about an accident if trained properly.

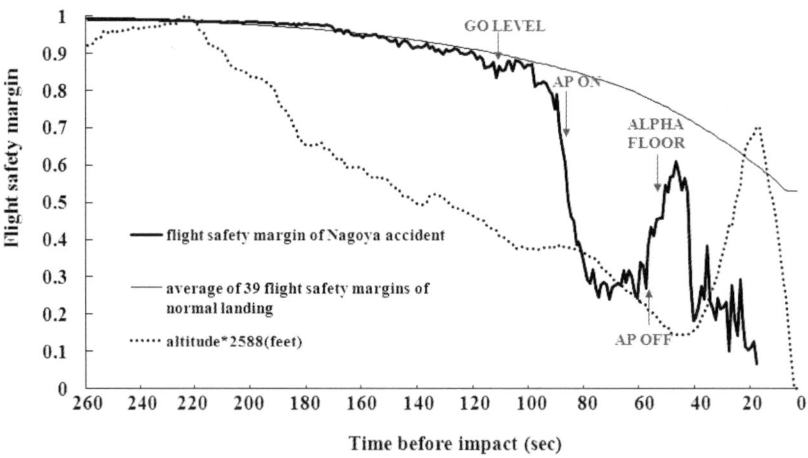

Figure 10.6 The flight safety margin in the Nagoya accident

The data from the accident report issued by the Japanese authorities were collected (Aircraft Accident Investigation Commission 1996). Starting from 2,400 feet to the ground – 258 seconds in total – the changes of all the corresponding situation parameters were put into the expert system. The calculated safety margin is given in Figure 10.6. One thing which should be noted is that the final safety margin for the accident is calculated down to 0.05, which is practically zero. After that, the accident would be unavoidable and could not theoretically be avoided by any pilot. After that, the calculation was stopped.

In Figure 10.6, the average of safety margins of normal flights was also given for comparison. From the very beginning, the calculated safety margin was relatively the same as that of a normal flight. At the time when the go-lever was triggered accidentally by the F/O, the safety margin started to decrease dramatically. At this moment, the automatic flight system started to execute the go-around order while the pilots still wanted to land. Because the efficiency of the trimmable horizontal stabilizer controlled by computer is larger than the elevator controlled by the pilots, the plane began to level off. After 13 seconds, the autopilot was also helping the plane to go around. The safety margin was compressed drastically. The speed of the airplane was further lowered. Up to this point, the pilots still could not convert the go-around mode to land mode. The conflict continued and became more serious with the enlarged difference between the elevator and the trimmable horizontal stabilizer. At 880 feet, the trimmable horizontal stabilizer reached –12.3 degrees while the elevator reached +8.5 degrees. The margin of safety left under this situation was only 0.25. Normally, the safety margin should be 0.82 at this altitude. At 700 feet, the autopilot switched off. The conflict between human and machine was partially relieved, resulting in an increase of the safety margin. However, the speed of the airplane was still decreasing. Further, the angle of attack was increased to 11.5 degrees. Unfortunately, the alpha-floor protection was activated at this time. The thrust was increased to maximum, resulting in a pitching movement to additionally raise the nose of the plane. The margin of safety was then further lowered and reached an unrecoverable situation. Finally, the zero margin was reached.

One thing here deserves to be noted: following the fluctuation of safety margin, the calculation was stopped when it reached 0.05, statistically zero. It was found that, at this point, the plane still had a speed of 31.5 knots and an altitude of 1,816 feet. In the meantime, the plane had already stalled and began to fall to the ground. Although the plane still had an altitude of 1,816 feet and 14 seconds before impact, the safety margin had theoretically already reached zero. It seems to be contradictory. However, this is consistent with the basic definition of the flight safety margin and the general perception of the pilots. The flight safety margin represents the room left for safe flight. There is little room for safe flight after a stall with so little airspeed so close to the ground. In addition, by definition, the flight safety margin was obtained from the inverse of the needed performance for recovery. If the calculated safety margin reaches zero, it meant that the needed performance for recovery was already at infinity, despite the fact that the height of the airplane was still at 1,816 feet. At this instant, the general perception of the local pilots is that the situation was already beyond any possibility of recovery by any human because the plane had already stalled and fell for one second. Although, during the survey, there were a certain number of pilots who believed that the plane was still recoverable from that situation, the average perception of all the subject pilots was that it was already not recoverable. Consequently, it is reasonable and acceptable that the flight safety margin did reach zero at 14 seconds before impact.

From the results shown above, the present theory and the corresponding expert system does preliminarily supply a number between zero and one which supposedly corresponds to the safety of flight. Flight safety margin theory does have the capability to reveal the change of perceived safety in a real accident.

4. The Da-Yuang Accident

Apparently very similar to the Nagoya case, the Da-Yuang accident originated from an entirely different beginning. The time history of the safety margin in this case is shown in Figure 10.7 with reference to the average of normal flights. From the very beginning, around 3,000 feet, the altitude of the plane was higher than required. The safety margin had already deviated and was lower than that of a normal flight. After the pilots recognized that the altitude was too high, the action they took was to further lower the flap to 20 degrees. This maneuver did not correct the error of a high approach. Consequently, the safety margin continued to be squeezed without any sign of recovery. Adversely, the lowering rate of the safety margin was gradually increased. At this instant, the safety margin was already close to 0.5, which is lower than the touchdown safety margin of a normal flight. As the speed decreased, the altitude was further lowered. When the airplane passed the outer marker, the altitude was still 1,000 feet higher than the normal altitude. This was not a stabilized approach. This mistake was never corrected throughout the flight's duration.

Several seconds later, it was found from the recorder that the angle of attack was lowered from 7 degrees to 5 degrees. The speed of the airplane was then gradually increased because of the drag reduction. The resulting safety margin recovered back a little bit and was close to 0.6, as shown in Figure 10.7. However, there was no sign of change in engine RPM. The increase in speed was not from the power of the plane. At the present situation, since the safety margin was as low as that at touchdown in a normal case, the airplane should commence the go-around immediately. However, the captain still wanted to land by telling the F/O to further bring down the flap to increase the drag. As expected, the speed of the plane was decreased further. The flight safety margin was already below 0.5 and advancing toward an even lower value. After the flap was lowered, the deviation in the safety margin from normal was further enlarged without any recovery. Even worse, to compensate for the decrease of the speed, the angle of attack was mistakenly increased. The safety margin was further compressed by this maneuver.

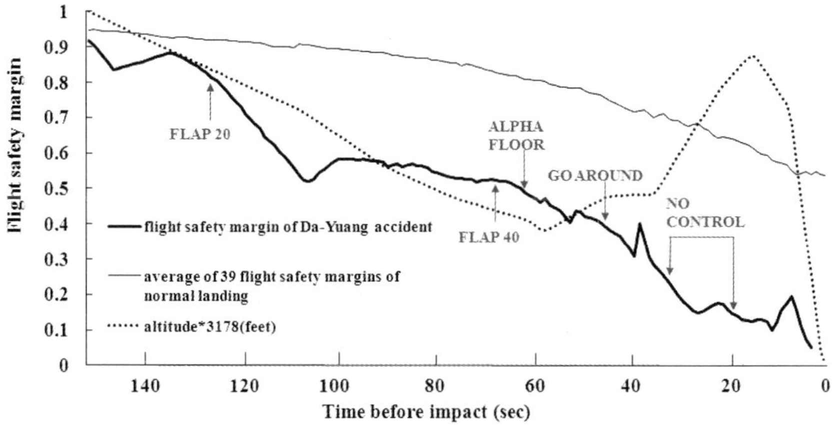

Figure 10.7 The flight safety margin in the Da-Yuang accident

Eventually, the alpha-floor protection was activated to save the plane from this dangerous situation. Because of the increase of pitch, the safety margin was squeezed further instead of being recovered by the activation of the alpha-floor protection. Noticing the rise of the nose, the crew pushed the control to lower it, resulting in the disengagement of the autopilot. The safety margin was slightly raised by this action. Since the speed of the plane was still decreasing, the resulting safety margin dropped again. At this instant, the safety margin was already below 0.4. After the crew decided to go around and released control to the automatic flight system to fly the plane (without knowing that it had been disconnected by their own action), there was no control, human or computer, of the plane for 11 seconds. At this time, the safety margin left was only 0.15. With an increasing angle of attack, the airplane stalled. No one could recover the plane since the safety margin was already almost zero. The calculation of the safety margin was stopped at the statistically zero point, 0.05.

From the variation of the safety margin, and the corresponding different flight situations, it is again proved that the present methodology can portray, to a certain extent, the variation of flight safety during the events leading up to an accident. Starting from an approach too high, safety was already threatened. Through the whole progression of this accident, this error was never corrected. Additionally, safety deteriorated further by the maneuvers of the crew. From the results shown above, the calculated safety margin was found to be able to roughly reflect the effect of these maneuvers on flight safety.

In this study, we have proposed a trial geometrical theory for flight safety, suggested by starting with Chinese ideographic thinking. The flight safety margin theory provides a very rough holistic measure instead of a sequential one for the safety of flights using a neural network to overcome the corresponding perception difficulty. From the results of the case study, some basic characteristics can be addressed. First, the proposed flight safety margin theory is scientific. Although the perceptions of the local pilots were included, the whole process was established scientifically and independent of individual opinion. Second, the present theory is quantitative. It can provide a measure for flight safety in different situations numerically. Third, this theory is reasonable and acceptable. According to the results, the present theory can provide a description, agreed upon by local pilots regarding the safety of real flight data, of normal operation to accident. Most importantly, the flight safety margin theory is obviously not a sequential-type theory. That is to say, it is surely possible to establish a scientific theory with Chinese thinking. Actually, it is not surprising since Chinese medicine has already existed for thousands of years.

With flight safety margin theory, the simulation capability can be basically established. It can also provide a tool to numerically evaluate the risk of human error. Combining these, the influence of human errors or other factors on the safety of flight can be estimated and compared. It could definitely be helpful to the management of flight safety since the defense can be placed reasonably at the needed spot. In addition, with the proposed theory, some kind of early warning capability could also be established. During flight, if the situation had already evolved into an unsafe cycle, yet still not severe enough to activate the warning system, the flight safety margin could provide an indicator showing that the situation should be noted. Concluding the present study, the main contribution of the present flight safety margin theory is the quantification of the perceived safety of flight. However, additional studies could be conducted for improvement of the system.

Chapter 11
A Harmonious Sky through Compatibility

Having paid a high price, Chinese culture is slowly adapting itself to Western aviation technology. Specifically through compliance with standard operating procedures and lowering the influence of Chinese authoritarianism, together with a change in the guanxi gradient between pilots on the flight deck, the progress is quite obvious. However, having digested these foreign (to the Chinese) ways of thinking, the advantages of a Chinese approach and value system will become more and more evident. With the Chinese-built COMAC C919 entering the market in the near future, the increasing Chinese influence in aviation is unavoidable, which means compatibility is even more important for a harmonious sky in the future.

11.1 Adaptation of Chinese Culture

Chinese culture has been trying very hard to adapt itself to Western science and technology for more than 200 years. Aviation represents the most difficult part of the whole evolution process since aviation technology, with its complexity of (computerized) flight management systems, air traffic management, airframes, propulsion, and many other technologies, is at the apex of Western science and technology. Having paid a high price, the adaptation to Western aviation technology was gradually improved. Before the end of the 20th century, the pattern of one major accident occurring in every four years was almost formed. Since then, the color of the sky in Taiwan has gradually changed from blood red to bright orange and is marching towards light blue – although there were still two major accidents after that, the primary cause of one of them was due to a foreign airline while the other was due to an error buried almost 30 years ago. Other than these, there has not been any major accident. The Aviation Safety Council of Taiwan published some statistics about the hull loss accidents from 1994 to 2012 (ASC 2012). Figure 11.1 shows the 10-year moving average accident rates of commercial turbojet and turbofan aircraft. One of the curves represents the hull loss accidents per million departures, while the other represents the per million flight hours. Although the numbers are still not good compared with international averages, the improvement is already quite obvious and steady.

The first explanation for this steady improvement is thought to be because the basic attitudes in the cockpit have already shifted closer to those of Western cultures. From the study in 2000 (Lu et al. 2000), some indications had surfaced. For example, question 66 of that questionnaire asked:

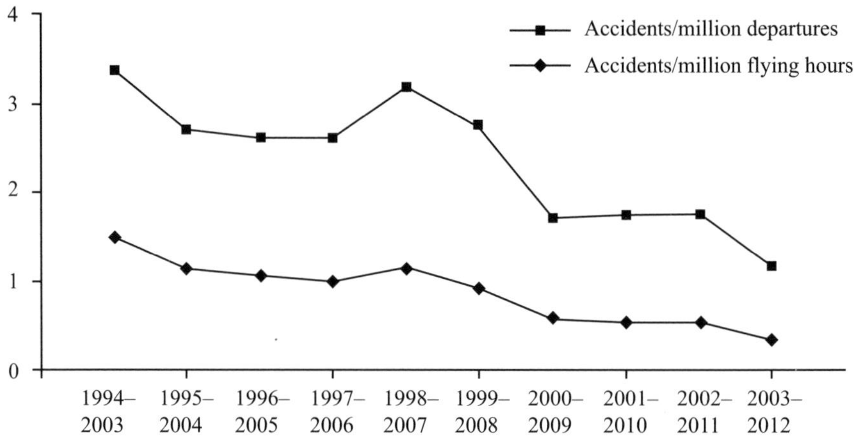

Figure 11.1 Ten-year moving average accident rates of commercial jet aircraft in Taiwan

> Q66: Which one of the above styles of leadership would you most prefer to work under?

This question asked about the styles of leadership from a more autocratic to a more democratic one. The attitude was almost the same for the local pilots with different ages. That means that this part of Chinese authoritarianism – auspicious expectation – remains almost the same as time passes. However, another question asked:

> Q67: In your organization, which style do you find yourself most often working under?

The score of the local pilots was 3.48 while the international average was 3.17 – more autocratic for Taiwanese pilots. The interesting thing in this observation was that a slight negative slope of the local scores varying with ages can be seen, which is found to be –0.074, as shown in Figure 11.2. It obviously showed that at least some of the younger pilots felt that the Chinese authoritarianism, at least the part of inscrutable power, was not as serious in the cockpit as recognized by the older pilots. It means that the younger pilots in Taiwan inherit less and less Chinese authoritarianism over time, although the speed of the change was quite slow.

The study in 2000 also included the influence of interpersonal relations on crew resource management. A part of that study was again executed in 2008 (Jing and Chen 2013). In the second study, these nine questions about the attitude toward different degrees of intimacy were asked again, and the underlying Chinese values were also investigated together with the guanxi gradient.

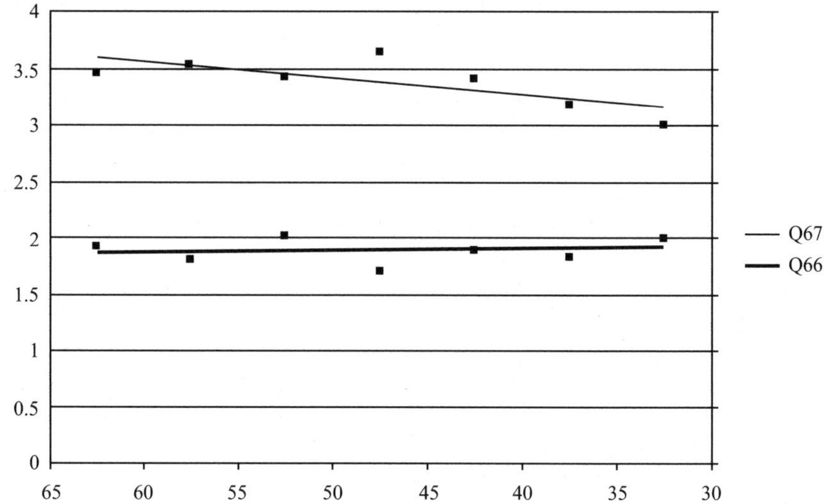

Figure 11.2 Variations of the pilots' perception about Chinese authoritarianism, by age

In 2000, the guanxi gradient was found to be –0.184, while it was –0.165 in 2008, as shown in Figure 11.3. The gradient was flattening, which means the attitude of Taiwanese pilots treating partners with different degrees of intimacy were becoming similar, i.e., not to particularly differentiate aliens from acquaintances.

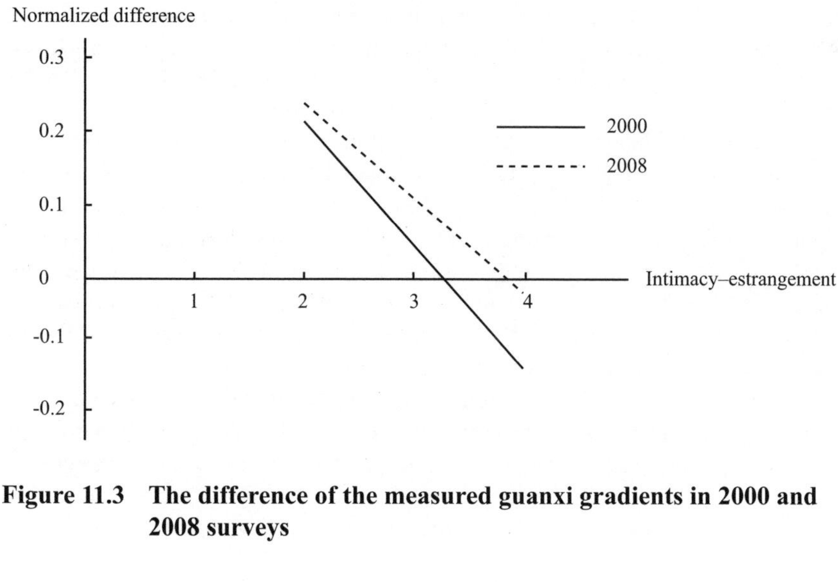

Figure 11.3 The difference of the measured guanxi gradients in 2000 and 2008 surveys

In conversations with local Taiwanese young people regarding foreigners nowadays, it is apparent almost everywhere. That is to say, the guanxi gradient of Taiwan does decrease.

The rate of flattening was calculated to be a little bit higher than zero, 0.002, per year. In the 2000, it was estimated to be 0.003 – the agreement is quite acceptable although the estimation from the linear extrapolation was a little bit higher than what was measured. Influenced by Western individualism, interpersonal relationships in Taiwan are becoming closer and closer to that of the West. However, the process takes quite a long time. According to the results, the evolution speed of the guanxi gradient is slower than expected in 2000, which represents the change of cultural traits in general is more difficult than we thought.

When one looks at variation in accident rates, some progress was revealed. Some of the factors affecting this included using English as the filter of culture in the cockpit, objective managerial tools such as the Flight Operation Quality Assurance system to enforce compliance with SOPs on the flight deck, and the recognition of the importance of flight safety for the general public. These, taken together with lowering the influence of Chinese authoritarianism and the guanxi gradient, led to accident rates declining. However, the process is quite slow. Hence, there must be something very fundamental that brought about the extreme low speed of adaptation of Chinese culture.

The most fundamental reason for the slow adaptation of Chinese culture to Western aviation technology is the mode of thinking. Having an ideographic thinking mode, the holistic or systemic nature of it creates problems and prevents Chinese people from thinking of the object separated from the subject; they very often mix themselves together with the object, which blurs the object in their thinking. This has been referred to by many Chinese scholars as the thought of the "oneness of heaven and humanity." Therefore, when Chinese people deal with their boss or subordinate, they not only simply think about the connection between the boss or the subordinate, but also involve some personal considerations. Furthermore, the Chinese also tend to complicate relationships into guanxi by integrating considerable personal classification and meaning when dealing with all kinds of interpersonal relationships. And the way that the Chinese think, in ideographic mode, does not go according to procedures, so Western aviation technology – which requires the Chinese to strictly adhere to procedures – is very alien. That is to say, when Chinese flight crews have to go according to procedures, they will face very high cultural pressure from inside and naturally want to bypass it. In normal conditions, one can use one's mind to overcome such a tendency. However, at stressful moments, earlier learning takes over, which would bring about the behaviors of not adhering to procedures and cause mistakes and even accidents.

The influence of mode of thinking not only exists in the adaptation to aviation technology, but also plays a crucial role when the Chinese culture faced the invasion of the Western civilization. Since the Opium War 170 years ago, Chinese culture has been forced to adapt to the world already shaped by Western civilization. The process was full of bitterness and tragedy. Western science and technology brought an unforgettable hurt and bitterness, including not only the humiliation from wars and the huge loss of lives and property, but also the frustration and failures when learning Western science and technology. What caused these frustrations and failures? When we dig more deeply into how the Chinese adapt to Western science and technology, we will be even more confused since most of the Western technologies that we are learning so diligently now were actually invented by earlier Chinese. Therefore, we must look deeply and seriously into our souls. Why did the Chinese invent gun powder and firearms, and then Westerners learned to use them to defeat the Chinese? Why did the Chinese invent the compass, yet it's Westerners who used it to achieve the great discoveries of European expansion? Why, in the Ming Dynasty, was China the only sea power in the world with the greatest armada in human history, yet hundreds of years later Western powers beat China with a much smaller fleet? Before the 15th century, Chinese technology was far ahead of the West. After the 15th century, Chinese technology still formed the foundation for the Renaissance and the corresponding rapid development of Western civilization. What could be the true reason to cause such a great turnaround between these two civilizations within just a few hundred years?

Chinese technology has flourished for centuries, as can be seen from the work of Joseph Needham (Needham 1973). However, there has never been such a theoretical framework like Western science formed in China. This is the well-known Needham problem. There are many explanations. Some people think it's due to the holistic thought of "oneness of heaven and humanity." Since Chinese thinking does not consider the object independent from the subject, the Chinese would not think about natural phenomena by excluding themselves, which means that the Chinese would always put some kind of connection between nature and themselves. Therefore, it is hard for the Chinese to think about natural phenomena independently, and thus it is not easy to simplify the target, very often leading to confusion and lack of clarity in interpreting the cause and effect of natural phenomena. On the other hand, it also brings about a utilitarian mentality for the Chinese; the Chinese would very naturally think, what kind of usefulness does the natural phenomenon have for me? Therefore, although China has never had any theoretical scientific construct as the West has, the Chinese do know how to make use of natural phenomena, which resulted in very advanced technology long ago in history.

The West inherited the Greek philosophical tradition and its corresponding formal logic, and was able to grasp and handle the tool of the linear mode of thinking. After the Renaissance, the West also discovered the possibility to find causal relationships through systematic experiment. By combining these two great discoveries, as Einstein said (Calaprice 2005), they built the only scientific system in human civilization. Therefore, they can use the advantages of this extremely sharp scientific inquiry system to improve and promote the technologies they have learned from China, and finally surpass the original inventor of those technologies, China.

When the Chinese again faced these Chinese technologies that had been fused into the Western scientific system, they could not recognize them – to such an extent that even after a very long time they still could not fully comprehend them. Of course, the core reason is obviously the lack of understanding of the basic thinking of Western science. Western science, based on Greek philosophy, is extremely alien to the Chinese way of thinking. Yet since science is the foundation of technology, the core issue of the Chinese adaptation to Western civilization is actually the issue of whether the Chinese can learn Western science. This is also the reason why "Mr. Science" would be the major theme demanded by top Chinese intellectuals in the New Cultural Movement, also known as the May Fourth Movement, which took place at Peking University in 1919.

Despite the huge efforts of Chinese intellectuals trying to learn Western science, even the most outstanding ones misunderstood Western science, simply because the difference between these two modes of thinking is just too fundamental. Take a very simple but crucial example. The physical theories created by Newton, Maxwell, Schrödinger, Einstein and others, including Newtonian mechanics and relativity, were understood by famous Chinese scholars educated at top universities such as Columbia University in the US as the objective, culture-free truth. However, the only object of these famous physical theories is to calculate results that can be compared with experiment (Dirac 1981) and have nothing to do with the objective truth. Further, as was said by Einstein: we are accustomed to regard as real those sense perceptions which are common to different individuals, and which therefore are, in a measure, impersonal. The natural sciences, and in particular, the most fundamental of them, physics, deal with such sense perceptions (Einstein 2005). It is quite clear that Western science is really not the objective truth as understood by those Chinese scholars. These terrible misunderstandings were passed down generation after generation through education up until today and still ongoing.

It has been nearly a hundred years since the May Fourth Movement held "Mr. Science" in such a high position. Though China has had significant progress in many areas of technological development, and has also achieved many remarkable feats in space technology, supercomputing, high-energy physics, civil and transportation technology, China still lags behind the West in scientific innovation. In science, no matter whether in the physical sciences or the biological sciences, there is still no one educated in China who has won the Nobel Prize, and the gap is even bigger for the social sciences. Those indications from different directions repeatedly tell us the same thing: the Chinese still cannot fully grasp the spirit of Western science. Why is this? The fundamental reason still comes from the huge difference in the modes of thinking of Western science and traditional Chinese thought (Jing 2012).

The mode of thinking of Western people is a kind of linear mode, emphasizing sequence based on the alphabetic system of writing: one-dimensional thinking. The traditional Chinese mode of thinking belongs to the ideographical mode, emphasizing a systematic view based on the Chinese character: two-dimensional thinking. It is obviously very hard for the Chinese with an ideographic mindset to learn Western science constructed by a linear tool, because it involves a transition of thoughts, not just the learning of knowledge. Moreover, when the

Chinese try to comprehend Western science with our traditional thinking without knowing the difference between the two modes, misunderstandings are almost inevitable. However, as long as the Chinese use their characters, their thinking mode will always be ideographic. And in the near future, it is impossible for the Chinese to give up using those characters, so Chinese thinking will definitely still be ideographic and there will always be an insurmountable obstacle between the Chinese and Western science.

However, there are of course advantages to ideographic thinking. Since the Chinese have a holistic view, they can think two-dimensionally instead of sequentially. When the Chinese think of the target object, their minds would automatically search in two dimensions for the most distinct features of the object as a whole, and can naturally find the invisible connections between different parts within it, then find the shortcut to reach the goal based on the Chinese utilitarianism nature, even when the information is incomplete. Of course, there is no standard operating procedure for this search to be based on. Yet the precondition of such a search is that the Chinese have to know enough about the system for them to find the most distinct features and the corresponding connections. Whenever the Chinese can grasp the features and connections, the two-dimensional systematic thinking can always supply more efficient ways to solve problems. In 1964, under the joint blockade enforced by the US and the Soviet Union, China successfully exploded an atomic bomb and became the fifth member of the nuclear club, despite being quite far behind in their level of science and technology. The same kind of thing happened again several years later. In 1971, even when China was still lagging behind the West in its industrial power, China independently launched a satellite into orbit, thus becoming the fifth country to enter into space. Even though China experienced a 10-year-long Cultural Revolution, which was considered to be a catastrophe to the society, China still sent a man into space with her own national capability in 2003, thus becoming the third country to execute manned space flight, after the US and the former Soviet Union. Now, in the current decade, as James Fallows describes in *China Airborne* (2012), China is set to join the United States and the European Union as a full partner in the global aviation community.

Before the Western invasion in the 19th century, Chinese civilization had survived for thousands of years; the ideographic mode of thinking has also been in operation for thousands of years. If there had been no invasion of Western ideas, China would continuously evolve and develop with her own path, steadily keeping the same direction since the surroundings did not change. However, in reality the surroundings did change, and were changed by a culture completely different from the Chinese. The kind of steady pace evolution to adapt gradually with the surroundings is no longer enough to meet this huge abrupt challenge. Chinese culture needs to go through a radical transformation to avoid extinction so as to achieve a new life, just like a phoenix reborn from the ashes of fire.

From a historical point of view, bringing Western science and technology to China might be similar to the arrival of printing, gunpowder, and the compass in the West. Extending back our historical imagination, Francis Bacon observed in 1620 that the Western world was being remade by printing, gunpowder, and the compass (Bacon 2003). They are all Chinese inventions. Printing created an educated class with mass literacy that could challenge the authority of popes and kings, and usher in a new form of political organization, the nation-state (Anderson 1983; Gellner 1983). Gunpowder made wars bloodier and gave decisive advantage to the state armies of Europe over less advanced tribes. The compass enabled the discovery and conquest of the New World. These are what we would call today "disruptive technologies," devices so different that they disrupted the stable European cosmos of the Middle Ages and resulted in many new things that brought huge influence to the world (Christensen 1997).

Just as those "foreign" devices of printing, gunpowder, and the compass – with comparative intellectual advantage – destabilized Europe, Western science and technology can also be seen as China's "disruptive technologies" for multiple reasons: they are alien and possess comparative intellectual advantage to China, they disrupted the Qing Dynasty and the corresponding long-lasting Dynasty Cycle of China, and directly or indirectly caused the birth of the Republic of China and the coming cultural reform lasting more than a hundred years. At the end of the Qing Dynasty, when China faced the far more advanced Western science and technology, the first thing to be destroyed was the traditional military system, which was quickly replaced by the Western military institution. The next to be destroyed was the educational system. The imperial examination system, which lasted for more than a thousand years, was dissolved and gave way to the Western university. The last to face disruption was the political system: the emperor system that centralized all power in one's hand built by the Qin Shihuang (259–210 BC), which sustained the middle kingdom for more than 2,000 years, finally collapsed and was replaced by the republican system from the West represented by the Republic of China. All of these revolutionary changes took place within 60 years. The invasion of Western science and technology in the 19th century caused an unimaginable change that had never happened in 3,000 years of Chinese history as perceived by most Chinese intellectuals.

These disruptive technologies are a kind of agent that dramatically adjusts internal structure when the culture is adapting to rapidly changing surroundings. Once the adjustment is completed, the culture can face its surroundings more efficiently. In the history of Chinese adaptation to Western science and technology, from the middle of the 19th century until the middle of the 20th century, after a hundred years of huge disruptions, Chinese culture has gradually entered into a stage of great stability. After the mid-20th century, Chinese ideographical thinking – having a tough time understanding Western science yet trying very hard to catch up with Western technologies – passed through the fall and stumble phases in 100 years and started to adjust its pace gradually to close the gap with Western science and technologies. After 60 years of difficult effort, civil aviation technology is the next area in which China tries to catch up. The fundamental reason is that the Chinese think they already understand Western aviation technology enough to be able to find the shortcuts in the system; that is to say, the Chinese can use the resources at hand to pursue a set target and gain enough differential advantage. The COMAC (Fallows 2012), represents the first step in China competing with the West – it has enough pieces of the civil aviation technology puzzle to be able to complete the whole picture.

11.2 Compatibility is a Two-way Street

In 2007, the Chinese government officially announced the start of the "big plane" project. The conglomerate Aviation Industry Corporation of China was established in 2008 to build the narrow-body, single-aisle, short- to medium-range C919, targeted to compete with the American B737 and European A320. The C919 plane will be produced by Commercial Aircraft Corporation of China (COMAC) in conjunction with a global network of component suppliers, expected to fly in 2014 and enter service in 2016.

Since the announcement, it has aroused heated discussions within the global aviation community, although until now the C919 is nothing more than a 20-foot-long model plane. However, no matter how small the model is, the C919 will definitely be a "big" plane for the world. Up to now, there have been many different kinds of comments. Despite the various opinions about the C919, Airbus and Boeing are always presented for comparison. Moreover, according to the planning, the engine and avionics will be Western-made. That is to say, the heart and brain of the C919 are all from the West, only the body belongs to the Chinese. Frankly, although the plane is "big," the first step of the C919 is truly small.

Even with such a humble start, the global sky has already been changed forever. The reason is simple. A great country such as China has a glorious, long history. It is impossible for the Chinese to let foreigners rule their skies forever. The C919 is a strategic decision of the Chinese government. Even if the C919 fails in the future, as the Y-10 did in the 1970s, China will still be struggling to fight for her independence in the sky. Moreover, even if the plane is mostly Western-made, it will definitely bear certain Chinese values within it, although these might be insignificant in the beginning. The Chinese influence in the sky will surely appear.

When the Western media talks about the C919, they normally use the word "threat" or "rival," or some other hostile words: such as the headline used by *USA Today* (October 15, 2009), "China wants to rival Boeing, Airbus with its C919 'big plane,'" and headlines from other media such as "The China Threat to Airbus and Boeing," or "Airbus and Boeing are building up big orders to squeeze out rivals but China's Comac could be a threat," or as commented in a research report by one of the consulting firms in the States:

> There is the very serious risk that by the time the C919 enters service (we think three years late is a good estimate), Airbus and Boeing are offering products that make this jet look obsolete. (quoted in Reuters 2012)

Obviously, Western interests view the C919 from a somewhat negative, dichotomous zero-sum competition angle, which is a very instinctive reaction from the linear mode of thinking of the West (Jing 2012). In their sub-consciousness, they always think that once the big plane produced by the Chinese is released, the huge advantages enjoyed by the West would finally be replaced and then disappear. However, not all competitions on earth are like this, be it in the biological context or cultural context. Although some competitions are zero-sum, there are even more that are in a co-existing symbiosis. It is possible that the rivalry of Western and Chinese aircraft manufacturers could benefit both, just as the rivalry of American pioneers such as the Wright Brothers and French pioneers such as Louis Blériot were mutually beneficial at the dawn of the Air Age.

The COMAC C919 is undoubtedly a competition of co-existence style. The engine of the C919 is built by the West. The backwardness of Chinese aero-engine technology remains the major bottleneck for the C919 program and the Chinese aviation industry as a whole. China does not have the technology to develop an aero-engine for the C919, which will thus be built around the aero-engine LEAP-X1C provided by the engine builder CFM International, a joint venture between GE Aviation of the US and Snecma of France. Besides, there are the avionics by Rockwell Collins and the nacelles and exhaust system by Nexcelle and CSeries cockpit from Bombardier. From what most of us know, only the fuselage and the corresponding materials will be built in China. From what's been shown above, in the near future, a large-scale change of these is not likely. Therefore, the release of the C919 actually also represents the interests of Western aviation technology.

So, What is the Real Meaning of the C919 to the Chinese?

In the long history of China, regardless of progress or lack of progress in the sciences, technological development has flourished. Before the 15th century, most of the technological innovations were more or less created in China. To China, what is hard is never the development of technology. As long as the knowledge of science starts to spread out to a certain extent, and the Chinese can learn and grasp comparatively well, China should be able to catch up with the West in technology. Therefore, the C919 actually represents an effort to keep abreast of or be on par with the West in terms of technological development, so as to let China bring benefit to its people after creating its own aviation technology.

However, to the Chinese, the development of the C919 is a problem of culture more than a problem of technology. Chinese civilization is the only ancient civilization that has been in continuous existence for millennia, a fact that most Chinese are extremely proud of. Every other dynastic civilization, from the Sumerians to the Babylonians to the Ashanti to the Maya, has come and gone. Although Chinese dynasties changed hands every 200 years in history on average, in the long history of China, struck by so many natural disasters and wars, with periods of prosperity and decline, Chinese civilization could always pass tests and overcome challenges. Chinese civilization endured and lives very strongly even today. Even when faced with the tremendous pressure of competition – which had never happened in 3,000 years – such as what Western science and technologies brought to the Chinese, Chinese civilization still overcame it and lives on. Looking toward the future, the strong and exuberant vitality of Chinese civilization will continue to lead the Chinese people to live on in the world, as has always been in the past 3,000 years.

The development of aviation technology represents only one aspect of the efforts of Chinese civilization to strive for survival on earth. To the Chinese, facing the harsh situation that Western science and technologies have almost dominated the development of human civilization, the Chinese have no other choice but to get involved proactively to avoid being eliminated. China can never allow herself to be absent from the civil aviation industry, because absence in this field means that China would become an outsider to the world's top technological development and it will eventually affect the continuity and prosperity of Chinese civilization. Therefore, the core meaning of the C919 for the Chinese is the inheritance and development of the Chinese civilization.

And, What Should Be the Appropriate Attitude of the West to the C919?

It is agreed generally that one of the most important fundamental values of Chinese culture is harmony. This was also confirmed in the CRM study (Jing and Chen 2013). Possessing an ideographic mode of thinking, the Chinese have a holistic, or systematic, view about almost everything, including competition with others. One key characteristic of the systematic view is the emphasis on dynamic interaction between the composing parts within it. To keep the system working positively, the Chinese believe that equilibrium should be maintained as the healthy interactions between different agents. If equilibrium is destroyed, the whole system will enter into a vicious spiral and finally break down, in which case everybody in the system will get hurt no matter who is a winner or loser, superior or inferior. Hence, the Chinese always try instinctively to avoid damaging the equilibrium in most situations.

However, equilibrium can sometimes be only an expectation, and it will always be shattered by the variations of any one component since the world will forever be changing. Therefore, we need a more proactive attitude to face changes. Equilibrium in some sense means no change, a somewhat passive attitude. Yet change is inevitable. A very proactive way of facing change is called compatibility. In Chinese minds, compatibility represents a kind of mentality which mutually admits and respects the existence and the corresponding rights of the other components in a system. For the dynamic interactions in a system, equilibrium means that every single entity plays its appropriate role; compatibility means to tolerate and compromise with the changes brought from each other and look for consistency of the whole system. Using concepts from Western philosophy to describe these concepts, equilibrium is about "being" while compatibility is about "becoming" for the components in a system. Both of them form a complete concept of Chinese harmony. The equilibrium and compatibility combined to form harmony moulds a very special characteristic of Chinese culture, sometimes called "self-restraint" by Western people.

Since 2008, the global aviation community has been changed by the Chinese COMAC. The previous equilibrium of the system established by Boeing and Airbus was thus shattered. Westerners see this change as a threat or rival. The Chinese did not conceive it in the same way. The world aviation community is by all means a huge system including China. Hence, when China develops the C919 to compete with Boeing and Airbus, holistic thinking still prevails. Subconsciously, the Chinese are still pursuing harmony; they don't want to destroy the present equilibrium of the global aviation community despite Western suggestions to the contrary. China is trying to implement compatibility to achieve the next equilibrium. The crucial point is whether Westerners view the C919 as a zero-sum competition by the Chinese, and the bottom line of that is whether Westerners accept China entering into the sky. However, Boeing and Airbus could not possibly be the only two powers in the sky forever; this isn't healthy for the global aviation community. Consequently, it will be more relevant for Westerners to treat the Chinese C919 as co-existing and co-prospering together, in which China is already implementing, instead of as a zero-sum competition. Therefore, compatibility is truly the key word for both China and particularly the West regarding development of the C919.

Compatibility cannot be achieved by any single member alone in a system since it is a two-way street. Roughly speaking, compatibility is an intermediate concept between union and struggle. Union represents working together and struggle means fighting each other, while compatibility is somewhere in between. In case of competition such as COMAC with Boeing and Airbus, some basic conflicts do exist. For example, transactions in the market will definitely be one of the hot spots, especially in the Chinese domestic market. However, even though conflict does exist, it is not zero-sum since the market will expand. At the worst case, COMAC cannot cooperate with Boeing and Airbus; this does not mean they have to fight. Above peace, it is definitely possible to find a way to interact through compromise. At least both of them can admit and respect the mutual existence of each other, which certainly cannot be achieved unilaterally.

Once compatibility is achieved, healthy interactions between COMAC, Boeing, and Airbus will bring about a prosperous global aviation community. This kind of prosperity has already been seen in free trade among the world's economies. The same is empirically true for the traffic in ideas: no advance in philosophy or the arts has come about except through a circulation of ideas. In the entire history of the natural and social sciences, the circulation of ideas has meant progress. The discoveries of chemistry and physics in the 17th and 18th centuries among scientists in France, Germany, England, and elsewhere in Europe, or the insights of economics and anthropology in the 18th and 19th centuries, were all achieved in a milieu of international communication and cooperation. The scientific community knows no geographic borders.

11.3 Toward a Harmonious Sky

Just as Einstein said, "Peace cannot be kept by force, it can only be achieved by understanding" (Calaprice 2005). If it takes mutual understanding to achieve peace, it would take yet more mutual effort and understanding for compatibility and cooperation. However, when it comes to mutual understanding, judging from the current situation, the West and the East are definitely not of equal status. Chinese people know quite a lot about the West, but the West knows comparatively little about China. For the better part of two centuries, Western societies have seen themselves as the model for all others since Western civilization had already dominated the world for hundreds of years, which makes Western people view China through a Western prism, and project Western history, culture, institutions, and values onto it. This situation is a lot like how China viewed the West in the 18th century. Having enjoyed superiority for too long, the ignorance was developed inside China. People automatically thought there was nothing to learn from the West.

More than 2,500 years ago, the great master of war, Sun Tzu, claimed: "Be aware yourself and your enemy, you will never be in fatal danger through battles." The Chinese paid a huge price in the 18th century for forgetting the lessons our ancestors had taught us. Of course, Westerners would not make the same mistakes as the Chinese did. Yet, history often repeats itself. If Westerners persist with their present mindset, we should not be surprised if they continue to get China wrong. However, that does not mean that the West needs to start speaking Chinese. Accept China as she has been throughout her long history rather than insisting on long-standing superiority, hubris, and presumption. It is thus necessary to improve this kind of uneven mutual understanding between the West and China and enhance mutual benefits of humanity including Westerners.

Once the balance of mutual understanding is achieved, compatibility and harmony will follow as well. Any single culture will know not to place its own values onto any other culture; the entirety of humanity will achieve a great status, as Mother Nature has shown: diversity. The essence of diversity is that nothing is universal. In this case, every single culture accepts all other cultures as they are. At that moment, the culture sphere of humans will be like that of the biosphere: diversity is the only fundamental value. This is also the ultimate wisdom and the top beauty this mother earth has given us. Evolution and progress proceed only through the interplay of a diversity of species, and a healthy, resilient ecosystem is one with a sufficient variety of species to withstand invasions and other environmental disturbances.

Before this ultimate goal of human kind can be achieved, the C919 could provide a humble trigger for it. China cannot be absent from the sky forever; this is the basic vitality of every single culture. And the West has to accept the third to appear in the sky, which would not necessarily mean that it will lead to a zero-sum competition. Since the sky is big enough, it will be able to accommodate multiple parties. Therefore, Boeing and Airbus, who have enjoyed a monopoly for a very long time, would have to learn to accept COMAC. A diverse sky would be just as healthy as the diverse biosphere because it will bring prosperity to all of us.

Our objective in this book has been to present – using methods (survey research) and concepts (CRM, cultural diversity) familiar in the West – a new perspective on this mutual adaptation. Adaptation always means mutual learning, whether by a community of collaborators or by multiple species in an ecosystem. Just as Europe learned navigation and print communication from China at the beginning of the modern era, and China is now learning modern technology from the West, so too the West can acquire an enlarged sense of cultural possibilities from Chinese thought styles and concepts of authority. The auspicious, benevolent dragon may not simply be in the cockpit of modern jetliners, but in the cockpit of the next century of cultural evolution, which can be beneficial for all mankind if it is based on mutual understanding. Let us start all of this from building a harmonious sky together.

References

Adorno, T.W., Frankel Brunswil, E., Levinson, D.J. and Sanford, R.N. (1950). *The Authoritarian Personality.* New York: Harper.

Aircraft Accident Investigation Commission (1996). *Aircraft Accident Investigation Report: China Airlines, Airbus Industrie A300B4-622R, B1816, Nagoya Airport, April 26, 1994*. Ministry of Transport, Japan.

Anderson, B. (1983). *Imagined Communities: Reflections on the Origins and Spread of Nationalism.* London: Verso.

Arnheim, R. (1969). *Visual Thinking*. Los Angeles, CA: The Regents of the University of California.

Aviation Safety Council (2012). *Taiwan Aviation Occurrence Statistics, 2003–2012*. Aviation Safety Council, Administration Yuan, Republic of China.

Bacon, F. (2003). *The New Organon* (Chinese translation). Beijing: The Commercial Press Ltd.

Batteau, A. (2010). Technological peripheralization. *Science, Technology, and Human Values*, 35(4), 554–74.

Bellah, Robert (1985). *Habits of the Heart: Individualism and Commitment in American Life*. Berkeley, CA: University of California Press.

Bo Yang (1977). *General History of China* (in Chinese). Taipei: Star Light Press.

Boeing Commercial Airplane Group (1994). *Flight Safety and Accident Investigation.* Workshop of Institute of Aeronautics and Astronautics, National Cheng Kung University, Tainan, Taiwan, R.O. China.

Braithwaite, G. and Caves, R. (1997). Airline safety – some lessons from Australia. *The Aeronautical Journal*, 1001, 29–32.

Calaprice, A. (2005). *The New Quotable Einstein.* Princeton, NJ: Princeton University Press.

Chang, H.-W. (1993). Word segmentation and sentence parsing in the reading process of Chinese (in Chinese). *Advances in the Study of Chinese Language Processing*, 1, 285–302. Department of Psychology, National Taiwan University.

Chang, L.-S. (2006). *Guanxi Gradient in Crew Resources Management* (in Chinese). M.S. Thesis, Institute of Aeronautics and Astronautics, National Cheng Kung University, Taiwan, R.O. China.

Chang, Y.L. (1942). Comments on the differences between the Chinese and Western cultures (in Chinese). *Thought and Era*, 11. Taipei.

Cheng, B.-S. (1995). Differentiated order and organizational behavior of Chinese (in Chinese). *Indigenous Psychological Study*. February, 142–219.

Chinese Social Science College (1983). *Bao-Ji Bei-Shau Hill* (in Chinese). Beijing: Artifacts Press.

Christensen, C. (1997). *The Innovator's Dilemma: When New Technologies Cause Great Firms to Fail.* Boston: Harvard Business School Press.

Chu, Y.-H. (1993). Power complex of Chinese (in Chinese). In Yang, K.-S. (ed.), *Development of Indigenous Psychology*. Taipei: Kwei-Guan Bookstore.

Chuan, Y.-C. and Yang, K.-S. (1997). Role-playing and cognition structure (in Chinese). *Indigenous Psychological Study*, June, 282–338.

Civil Aeronautics Administration (1997). *Aircraft Accident Report, TransAsia Airways, ATR72-200, B-22717, Tao-Yuang, January 30, 1995* (in Chinese). Taiwan, R.O. China: MOTC.

Civil Aeronautics Administration (2000). *Aircraft Accident Report, China Airlines, Airbus A300B4-622R, B-1814, Da-Yuang, Tao-Yuang, February 16, 1998*. Taiwan, R.O. China: MOTC.

Civil Aeronautics Administration (2001). *Aircraft Accident Report, Formosa Airlines, SAAB-340, B-12255, March 18, 1998*. Taiwan, R.O. China: MOTC.

Civil Aeronautics Administration (2007). *Safety Management System, AC120-32B*. Taiwan, R.O. China: MOTC.

DeFrancis, John (1984). *The Chinese Language: Fact and Fantasy*. Honolulu. University of Hawaii Press.

Dirac, P.A.M. (1981). *The Principles of Quantum Mechanics*. Oxford: Oxford University Press.

Douglas, M. and Wildavsky, A. (1982). *Risk and Culture: An Essay on the Selection of Technological and Environmental Dangers*. Berkeley, CA: University of California Press.

Einstein, A. (2005). *The Meaning of Relativity*. Taipei: The Commercial Press Ltd.

Embry-Riddle Aeronautical University (1999). *Flight Operations Manual*. Daytona Beach, FL: Embry-Riddle Aeronautical University.

Fallows, J. (2012). *China Airborne.* New York: Pantheon Books.

Federal Aviation Administration (2006). *Introduction to Safety Management System for Air Operators*. Advisory Circular 120–92. Washington, D.C.: Federal Aviation Administration

Fei, X.-T. (1947). *The Native China* (in Chinese). Hong Kong: San-Lian Books Company.

Fischer, S.R. (2001). *A History of Writing*. London: Reaktion Books.

Frederiksen, J.R. (1981). *Interactive Processes in Reading*. Cambridge: Lawrence Erlbaum Associates.

Gagné, E.D., Walker Yekovich, C. and Yekovich, F.R. (1993). *The Cognitive Psychology of School Learning* (in Chinese, trans. Yue, X.-P.). Taipei: Yuan-Liou Press.

Gan-Su Museum (1979). *Gan-Su Painted Pottery* (in Chinese). Beijing: Artifacts Press.

Gau, C.-S. and Chen, C.-S. (1989). Social order in the operation of enterprise in Taiwan – favor, relation, and the law (in Chinese). *Society and Economics*, 3(4), 151–65.

Gellner, E. (1983). *Nations and Nationalism*. Ithaca, NY: Cornell University Press.

Giarratano, J. and Riley, G. (1994). *Expert System – Principles and Programming*. Boston, MA: PWS Publishing Company.

Gras, A., Moricot, C., Poirot-Delpech, S.L. and Scardigli, V. (1994). *Faced with Automation: The Pilot, the Controller, and the Engineer.* Paris: Publications de la Sorbonne.

Harris, D. (2011). *Human Performance on the Flight Deck*. Farnham: Ashgate.

Harrison, L.E. (2006). *The Central Liberal Truth: How Politics Can Change a Culture and Save It from Itself.* Oxford: Oxford University Press.

Hawkins, F.H. (1987). *Human Factors in Flight*. Brookfield, Vermont: Gower Publishing.

Haykin, S. (1999). *Neural Network – A Comprehensive Foundation*. Upper Saddle River, NJ: Prentice Hall.

Heinrich, H.W. (1950). *Industrial Accident Prevention*. New York: McGraw-Hill.

Helmreich, R.L., Klinect, J.R. and Wilhelm, J.A. (1999). Models of threat, error, and CRM in flight operation. *Proceedings of the Tenth International Symposium on Aviation Psychology*. Columbus, OH.

Helmreich, R.L. and Merritt, A.C. (1998). *Culture at Work in Aviation and Medicine: National, Organizational and Professional Influences*. Aldershot: Ashgate Publishing Limited.

Ho, D.Y.F. (1998). Interpersonal relationships and relationship dominance: an analysis based on methodological relationalism. *Asian Journal of Social Psychology*, 1, 1–16.

Ho, Y.-H. and Chen, S.-C. (1991). Relation orientation: finding solutions for methodology of Chinese social psychology (in Chinese). In Yang, K.-S. and Huang, K.-K. (eds), *Psychology and Behavior of Chinese*. Taipei: Kuei-Kuwan Books Company.

Hofstede, G. (1980). *Culture's Consequences: International Differences in Work-related Values*. Beverly Hills, CA: Sage.

Hsu, C.Y. (1990). *History of Western Chou* (in Chinese). Taipei: Lian-Jin Books.

Hsu, F.L.K. (1970). *Americans and Chinese: Two Ways of Life*. New York: Natural History Press.

Hsu, F.L.K. (1981). *Americans and Chinese – Passage to Differences*. Honolulu: The University Press of Hawaii.

Hsu, F.L.K. (2000). *The Study of Literate Civilizations* (in Chinese). Taipei: Nan Tian Books.

Hsu, S.-S. (1998). *Words-Instead-of-Characters as Presentation Units in Chinese Text (II): Layout Parameter Determination* (in Chinese). Taipei: Science and Technology Information Center, National Science Council.

Huang, K.-K. (1995). The favor and relations of Chinese (in Chinese). In Wen, C.-I. and Shaw, S.H. (eds), *Chinese Concept and Behavior.* Taipei: Ju-Liou Books Company.

Huang, K.-K. (1998). *The Game of Power of Chinese* (in Chinese). Taipei: Ju-Liou Books Company.

Jer Jian Archeological Research Institute (1988). Jer Jian Yu Hen excavation report (in Chinese). *Artifacts*, 1.

Jing, G.-T. and Liou, C.-F. (1994). *Flourish and Crisis – On the Super Stable Structure of Chinese Society* (in Chinese). Taipei: Storm & Stress Publishing Co.

Jing, H.-S. (1998). *A Native Thinking for Flight Safety in Taiwan* (in Chinese). Conference for the Review and Promotion of Flight Safety in Taiwan, Civil Aviation Authority, Department of Transportation, R.O. China.

Jing, H.-S. (2004). Differentiated order in the cockpit. *Human Factors and Aerospace Safety*, 4(2), 131–43.

Jing, H.-S. (2012). *The Critique of Western Philosophy* (in Chinese). Taipei: Science Monthly Publishing Co.

Jing, H.-S. and Chang, L.-S. (2006). Guanxi gradient and flight safety. *Human Factors and Aerospace Safety*, 6(1), 17–34.

Jing, H.-S. and Chen, B. (2013). *The Influence of Guanxi Gradient on Crew Resources Management and Values in the Cockpit.* 15th International Conference on Human-Computer Interaction, Nevada, USA, July 21–6.

Jing, H.-S. and Lee, J.-H. (2001). Grouping pattern of Chinese and its influence on the understanding of the flight operation manual (in Chinese). *Civil Aviation Journal Quarterly*, 3(4), 43–58.

Jing, H.-S., Lu, C.J. and Peng, S.-J. (2001). Culture, authoritarianism and commercial aircraft accidents. *Human Factors and Aerospace Safety*, 1(4), 341–59.

Jing, H.-S., Lu, P.-J., Yong, K. and Wang, H.-C. (2002). The dragon in the cockpit: the faces of Chinese authoritarianism. *Human Factors and Aerospace Safety*, 2(3), 257–75.

Job, M. (1994). *Air Disaster*. Fyshwick, Australia: Aerospace Publications Ltd.

Just, M.A. and Carpenter, P.A. (1987). *The Psychology of Reading and Language Comprehension*. Boston, MA: Allyn & Bacon.

King, Y.-K. (1992). Construction of relation and network: A sociological interpretation (in Chinese). *Twenty First Century*, August, 143–57.

Kou, H.M. (1915). Spirit of the Chinese People. *Daily News, Peking.*

Lee, Z.H. (2008). *Essays on Chinese Traditional Thoughts* (in Chinese). Tien-jin, China: Institute of Social Science Press.

Liang, S.M. (1963). *The Element of Chinese Culture* (in Chinese). Taipei: Chi-Chen Books Company.

Liu, J.-X. and Yang, J.-R. (1999). *Dragon and Chinese Culture* (in Chinese). Beijing: People Press.

Logan, R.K. (1986). *Alphabet Effect – The Impact of the Phonetic Alphabet on the Development of Western Civilization.* New York: William Morrow and Company, Inc.

Lu, P.-J., Wang, S.-E., Co, H.-J., Jing, H.-S., Yuan, S.-F. and Dai, M. (2000). *Corporate Resource Management and Corporate Culture Research Report* (in Chinese). Taipei: Aviation Safety Council.

Luria, A.R. (1973). *The Working Brain: An Introduction to Neuropsychology.* London: Penguin Books.

Man, J. (2000). *Alpha Beta: How Our Alphabet Shaped the Western World.* London: Headline Book Publishing.

McCormick, B.W. and Papadakis, M.P. (1996). *Aircraft Accident Reconstruction and Litigation.* Tuscon, AZ: Lawyers & Judges Publishing Company.

Miller, G.A. (1956). The magical number seven, plus or minus two: some limits on our capacity for processing information. *Psychological Review*, 63, 81–97.

Mumaw, R., Sarter, N. and Wickens, C. (2001). Analysis of Pilots' Monitoring and Performance on an Automated Flight Deck. *Proceedings of the 11th International Symposium for Aviation Psychology*. Columbus, OH, USA.

National Research Council (1997). *Aviation Safety and Pilot Control – Understanding and Preventing Unfavorable Pilot-Vehicle Interactions.* Washington, D.C.: National Academy Press.

Needham, J. (1973). *Science of Civilisation in China* (Chinese translation). Taipei: The Commercial Press, Ltd.

Ortner, S.B. (1973). On Key Symbols. *American Anthropologist*, New Series, 75(3), 1338–46.

Perez, A. (2002). The death spiral in CFIT accidents. Unpublished MA thesis. Wayne State University, Detroit, Michigan.

Perrow, C. (1984). *Normal Accident: Living with High-Risk Technologies.* New York: Basic Books.

Polanyi, M. (1967). *The Tacit Dimension.* Garden City, NY: Anchor Books.

Pu-Yang Artifacts Management Council (1988). Henan Pu-Yang West Water Hill site archeological report (in Chinese). *Artifacts*, 3.

Pye, L.W. (1968). *The Spirit of Chinese Politics: The Psychocultural Study of the Authority Crisis in Political Development.* Cambridge, MA: MIT Press.

Pye, L.W. (1981). *The Dynamics of Chinese Politics.* Cambridge, MA: Oelgeschlager, Gunn and Han.

Rasmussen, J. (1997). Risk management in a dynamic society: a modelling problem. *Safety Science*, 27(2/3), 183–213.

Rasmussen, J. and Jensen, A. (1974). Mental procedures in real-life tasks: a case study of electronic troubleshooting. *Ergonomics*, 17, 293–307.

Reason, J. (1990). *Human Error*. New York: Cambridge University Press.

Reuters. (2012). Wing cracks, other flaws delay China jet manufacture. Reuters news wire, June 8, 2012. http://in.reuters.com/article/2012/06/08/uk-airlines-china-comac-idINLNE85700Z20120608 (downloaded August 9, 2014).

Rockhill, W.W. (1897). Diplomatic missions to the court of China: the kotow question I. *The American Historical Review*, 2(3), 427–42.

Rockhill, W.W. (1897). Diplomatic missions to the court of China: the kotow question II. *The American Historical Review*, 2(4), 627–43.

Sarter, N. and Woods, D. (1995). From tool to agent: the evolution of (cockpit) automation and its impact on human-machine coordination. *Proceedings of the Human Factors and Ergonomics Society*, 1, 79–83.

Sheng, C.-S. (2006). *Flight Safety Margin – A Geometrical Point of View for Flight Safety* (in Chinese). MS thesis. Institute of Aeronautics and Astronautics, National Cheng Kung University, Taiwan, R.O. China.

Strauch, B. (2002). *Investigating Human Error: Incidents, Accidents, and Complex Systems*. Aldershot: Ashgate Publishing.

Sun, S.-D. (1984). San Xin Ta La Horng-Shan Jade Dragon Study (in Chinese). *Artifacts*, 6.

Tessier, C. and Dehais, F. (2012). Authority management and conflict solving in human-machine systems. *Aerospace Lab Journal*, 4(May), 1–10.

Trompenaars, F. and Hampden-Turner, C. (1993). *The Seven Cultures of Capitalism: Value Systems for Creating Wealth in the United States, Japan, Germany, France, Britain, Sweden, and the Netherlands. i.* New York: Doubleday.

Trompenaars, F. and Hampden-Turner, C. (1998). *Riding the Waves of Culture: Understanding Diversity in Global Business*. New York: McGraw-Hill.

Tylor, E.B. (1871). *Primitive Culture: Researches into the Development of Mythology, Philosophy, Religion, Language, Art, and Custom*. London: J. Murray.

Tzeng, W.-S. (1972). Personality of Chinese from development point of view (in Chinese). In Yang, K.-S. (ed.), *Chinese Personality*. Taipei: Kwie-Guan Bookstore.

Watanabe, M. (1998). *Styles of Reasoning in Japan and the United States: Logic of Education in Two Cultures*. San Francisco, CA: American Sociological Association.

Weber, M. (1964). *The Religion of China*. New York: Free Press.

Wei, C.-T. (1972). Analysis of traditional ideal Chinese personality (in Chinese). In Yang, K.-S. (ed.), *Chinese Personality*. Taipei: Kwei-Guan Bookstore.

Weick, K. (1990). The vulnerable system: an analysis of the Tenerife air disaster. *Journal of Management*, 16(3), 571–93.

Wen, C.-I. (1972). Personality of Chinese from value point of view (in Chinese). In Yang, K.-S. (ed.), *Chinese Personality*. Taipei: Kwei-Guan Bookstore.

Wen, I.-D. (1948). *Dragon and Phoenix* (in Chinese). Beijing: Kai-Nin Bookstore.

Wood, R.H. and Sweginnis, R.W. (1995). *Aircraft Accident Investigation*. Casper, WY: Endeavor Books.

Yang, C.-F. (1993). Are Chinese really collectivists? (in Chinese). In Yang, K.-S. (ed.), *Chinese Value System: Social Science Point of View*. Taipei: Kwei-Kuwan Books.

Yang, C.-F. (2001a). *The Interpersonal Relation, Sentiment and Trust of the Chinese People* (in Chinese). Taipei: Yuan-Liou Publishing Co.

Yang, C.-F. (2001b). *How to Understand the Chinese* (in Chinese). Taipei: Yuan Liou Press.

Yang, H.-M. (1998). The influence of adjustment in space size between and within words on Chinese reading (in Chinese). *Academic Journal of National Taiwan University*, 31, 303–26, Dept. of Psychology, National Taiwan University, R.O. China.

Yang, K.-S. (1993). Social orientation of Chinese: social interaction point of view (in Chinese). In Yang, K.-S. and Tu, A.-B. (eds), *The Psychology and Behavior of Chinese – Methodology*. Taipei: Kwei-Kuwan Books Company.

Yang, L.S. (1957). The concept of "pao" as a basis for social relation in China. In Fairbank, J.K. (ed.), *Chinese Thought and Institutions*. Chicago, IL: University of Chicago Press, 291–309.

Yang, Y.-I. (2000). Insiders: a case study about classification of Chinese human relation (in Chinese). *Indigenous Psychological Study*, June, 277–316.

Yu, Y.-S. (1984). *Modern Meaning of Chinese Culture from Value System* (in Chinese). Taipei: Times Culture Publications.

Index

Note: **bold** page numbers indicate figures and tables.